Of Clocks and Time

Of Clocks and Time

Lutz Hüwel
Wesleyan University, Middletown, USA

Morgan & Claypool Publishers

Rights & Permissions
To obtain permission to re-use copyrighted material from Morgan & Claypool Publishers, please contact info@morganclaypool.com.

ISBN   978-1-6817-4096-6 (ebook)
ISBN   978-1-6817-4032-4 (print)
ISBN   978-1-6817-4224-3 (mobi)

DOI   10.1088/978-1-6817-4096-6

Version: 20180401

IOP Concise Physics
ISSN 2053-2571 (online)
ISSN 2054-7307 (print)

A Morgan & Claypool publication as part of IOP Concise Physics
Published by Morgan & Claypool Publishers, 1210 Fifth Avenue, Suite 250, San Rafael, CA, 94901, USA

IOP Publishing, Temple Circus, Temple Way, Bristol BS1 6HG, UK

To Ruth.

Contents

Preface

With this book, I invite you to join me in an exploration of ideas concerned with time and the tools used to measure it. It seems reasonable if not logically required to define a quantity first before one can try to measure it. Therefore it is puzzling that *time* can be measured with astonishing precision[1] while a rigorous definition of what time actually *is*—if it exists at all—remains elusive. In physics, time is simply identified as that aspect of nature that certain instruments measure. These instruments are commonly called *clocks*. However, what constitutes a clock is left vague lest circular logic needs to be invoked ('a clock is a device to measure time'). While this state of affairs is not very satisfactory, we may take comfort in St Augustine's struggle about the nature of time: 'What, then, is time? If no one asks me, I know what it is. If I wish to explain it to those who ask me, I do not know' [1]. Some things are just plain difficult. But the conundrum has neither prevented improvements, vast improvements at that, in clock technology nor the development of sophisticated concepts and theories about time. In physics, we can sometimes get away with pragmatic approaches. The hard part is then left to the philosophers [2]. My approach is simply to describe a variety of observable, measurable aspects related to time with the hope that this multifaceted collection can point to a useful—even if approximate—way to understand time.

The story I want to tell begins with a few natural phenomena that relentlessly mark both linear change and cyclic repetitiveness: day and night, the lunar cycle, and the course of the Sun and the stars in the sky (chapter 1—Days, months and years). In this chapter, astronomical observations are summarized, as well as the methods with which they have been obtained and how modern physics explains them. Many people living in different cultures and eras have contributed to this accomplishment —it was a truly global endeavor. Calendars of all types are testimony to this effort, as are ancient monuments and artifacts such as the Kukulkan pyramid, built centuries ago in the jungles of Chichen Itza, or the Antikythera mechanism, rediscovered almost 2000 years after it was lost in the waters of the Aegean Sea. While its manifold beginnings are obscure, the enterprise of observing the heavens has produced well-documented and astonishing achievements—among them the Copernican revolution and Newton's foundation of modern physics. Seeing clearer and further is and was only possible because others have been on the lookout before. It turns out that Newton in particular will be a constant companion helping us along the way to gain insight—including in those instances where he turned out to be wrong.

---

[1] Currently (fall 2017), the precision of the official US time-keeping clock, the NIST-F1 cesium fountain clock is about one part in 3×10^{16} or one second in about 100 million years (www.nist.gov/pml/div688/grp50/primary-frequency-standards.cfm). Since clocks with far greater precision are in development, it seems likely that at least for the foreseeable future time will remain much more precisely measurable than any other quantity. The above quoted measurement uncertainty is roughly equivalent to specifying the US national debt to within one cent—given the debt's rapid change, an impossible and fruitless task. How such a precision can be obtained and why it can be useful will hopefully become clear in later chapters.

The tale will continue with the effort to divide the natural temporal units of year, month and day into finer and finer parts. Here begins the history of clocks in the more common usage of the word, e.g. sundial, hourglass and mechanical clock (chapter 2—Hours, minutes and seconds). Real events usually do not unfold in a neat, linear fashion; stories recounting real events can be streamlined a bit, but in the end, they too are often more similar to a quilt than to beads on a string. The history of mechanical clocks, for example, starts independently and well before the narrative of annual and daily cycles came to a preliminary conclusion, in chapter 1, with Kepler's empirical laws of planetary motion and Newton's explanation of those laws in his universal theory of gravity. Further complicating the relation between these two strands of the story of time-telling is the fact that the improvement of mechanical clocks has helped to perfect celestial observations tremendously. Chapter 2 also contains a discussion of a few basic questions related to clocks in general. What are the minimum ingredients? What does it take to build a perfect clock? How do you measure the approach to perfection? Simple mechanical systems, such as the pendulum and a bead attached to a coiled spring, will serve us well as toy models. Their behavior can be analyzed quantitatively (thanks, Isaac), yielding benchmarks and useful parameters for comparison with other, more realistic clocks.

In the 19th and 20th centuries, the accelerated pace of technological development, commerce and other social interactions meant that even a fleeting second was found to be too long for the need of precise and accurate measurements. Replacing the ticking of man-made oscillators with the resonant vibrations of small quartz crystals was a breakthrough in the precision and stability of clocks. Shortly after the invention of the quartz clock, the time-keeping quartz oscillator was in turn replaced by the even more exact and steady vibrations of atoms. Since then atomic clocks have been fine-tuned to become machines of awesome precision. Chapter 3 (From milliseconds to attoseconds: is there a limit?) follows this thread of the story and explores the current limits with which rates of change of physical quantities can be ascertained. The chapter also contains a rudimentary synopsis of atomic physics as it pertains to the functioning of atomic clocks. In addition, I will try to shed further light on the notion of clock precision and stability.

The development of atomic clocks unfolded in the context of two separate revolutions in physics that have fundamentally altered our view of the world. Chapter 3 must invoke quantum theory in order to explain the inner workings of atomic clocks. Much more than a manual for atomic and molecular structure and behavior, this theory offers new paradigms flowing from an inherently statistical nature of change and a deep interconnectedness between different parts of a given system, including that between observed and observer. All this leads to sometimes counterintuitive and often difficult-to-grasp ways in which change happens (or does not) or is observed (or is not). The second transformation comes in two parts, Einstein's two theories of special and general relativity. Both reveal the profoundly different ways in which space and time manifest themselves to observers in different states of motion. Chapter 4 (Space and time forever entwined) summarizes the basic concepts of these theories—severely abbreviated to fit the purpose at hand. As has been pointed out many times by others, Einstein's theory of relativity might be called with equal justification, 'a theory of the absolute'. Simultaneity and the duration of temporal intervals do indeed depend on the state of the observer. However, the speed of light and a suitably formulated space–time interval between any two events are absolute, agreed upon by all observers. How this comes about and what it means for our understanding of time and its measurement will be the main emphasis of this part of the story.

Periodic transformations and circular time, irreversible changes and linear time—these two themes provide a clear distinction between notions of time in different cultures. They are also at the core of two different strands of ideas in physics. For the most part, the story up to this point has been about circular time and things that happen periodically. But even a set of atomic fountain clocks, at one moment synchronized to the highest possible precision, will run out of tune eventually. Initial coherence will be lost and the arrow of time will assert itself: there is a direction of time. One truly stochastic, unidirectional process—radioactive decay—has been used to serve as an altogether different type of clock. Chapter 5 (Deep time or getting old) explores the underlying physics and working principles of this approach to measuring time. We will see that such 'clocks' are well suited to determine the age of both natural and manmade things from a few hundred up to billions of years. Old as some rocks on Earth are, there are older things. The birth of Earth is not the beginning of time. That is an entirely different story that has been pieced together by the marriage of astronomical observation with physical theories. A grand, consistent image has emerged that is known somewhat frivolously as the Big Bang theory. Chapter 6 (From beginning to end), the final one, traces the basic ideas and observations of this aspect of telling time, which, with pleasing symmetry, connects the narrative back to the starting chapter.

One last warning before I begin: the linear sequence of the narrative belies the multi-dimensional connections between the topic that each chapter tries to focus on. Furthermore, there was never a neat linear progression of our thinking, dare I say understanding, of time—or any subject for that matter. For this reason, I will not always be able to follow a straightforward trajectory in my story. Truth be told, I also like to digress on occasion.

So then, what is time?[2]

We all know time can fly. And in a song by The Alan Parson Project, time 'keeps flowing like a river'. I like that metaphor: on occasion there is but a trickle and water-time hardly moves, but soon enough it rushes forward with great speed and force. Then again, the comparison has its limits—like all analogies do. Water flows, but time is not a substance. So what is flying when time flies? This is a difficult question. Maybe so difficult that all we can do is use analogies, after all. Still, many valiant attempts have been made to find an answer to the question of what time is. I

---

[2] A version of 'So then, what is time?' was published in 2012 on the Wesleyan website [http://passingtime.blogs. wesleyan.edu/what_is_time/].

marvel at St Augustine's struggle with the concept of time in his *Confessions* [1]. In the end, though, he is forced to admit: 'What, then, is time? If no one asks me, I know what it is. If I wish to explain it to him who ask me, I do not know'. In physics, the difficulty of coming to terms with what time is has been cleverly sidestepped by using an operational definition: time is what a clock says it is. Question: what is a clock? Answer: a device to tell time. While this wordplay is of course facetious, even careful definitions of what constitutes a clock run into the risk of circular logic. St Augustine really has a point! Nevertheless, the seemingly limiting and vulnerable approach to restrict discussion of time to measurable aspects has yielded fascinating and deep insights into the nature of time. Take Einstein. As a starting point for the theory of relativity, he substitutes 'position of the small hand of my clock' for 'time'. With that simple trick and the constancy of the speed of light—assumed to be perfect, measured to be true for all circumstances tested so far—he manages to predict properties of time, all verified by now, that a hundred years later remain strange and somewhat disturbing. Events judged to be simultaneous by one observer have a finite span of time between them for others. Temporal order is not absolute— the same event that occurred in the past for me can still lie in the future for you. Time can be stretched and compressed—two identical clocks can tick at their own, distinctly different beat. There is not just one absolute and universal flow of time as Newton had envisioned, but a multitude of time rivers. Another facet having to do with the notion of flowing time intrigues me. Is time continuous or does a smallest, indivisible unit exist? No change is truly experienced as continuous movement, but rather as a sequence of still images. But that does not imply time itself is discontinuous. The quest for ever more exactitude has given us clocks that chop a nanosecond into a million pieces; no limit seems to be in sight. It is quite astonishing how simple acts of observation and reasoning have contributed so much to what we know about time. It is likewise remarkable with what delicate precision we can measure time while a true understanding still eludes us.

Lutz Hüwel, Wesleyan University

References

[1] Augustine 1955 *Confessions and Enchiridion* (The library of Christian Classics vol VII) transl. and ed A C Outler (Philadelphia, PA: The Westminster Press) Bk 11, Ch. XIV, Para. 17 https://sourcebooks.fordham.edu/halsall/basis/confessions-bod.asp
[2] *Internet Encyclopedia of Philosophy* www.iep.utm.edu/time

Acknowledgments

When does a book project start? In many cases, I suspect, there is no definitive answer. For my book *On Clocks and Time* I know that there is none. Nevertheless, several occasions and events in the past have something to do with the fact that the book came into existence. Always, someone or some people were involved and this short acknowledgment is my attempt to give thanks to them.

First of all, I have to admit that for the longest time, the thought of writing this book, *any* book, never crossed my mind. Nevertheless, already many years ago and unbeknownst to me, the seeds were sown. In hindsight, it all began with a course for non-science majors—*It's About Time*—that I developed and taught at Wesleyan University or simply Wesleyan, as we refer to our academic home here in Middletown, Connecticut. From the beginning, I used the modular approach that is reflected in this book. Certainly, some of the material was already present in the first incarnation of the course which occurred soon after a sabbatical semester at JILA in Boulder, Colorado. That sequence is significant. A public lecture on time measurement, a topic very much at home in Boulder, intrigued me to the point of getting serious with my general education course. Thus, my first thanks belong to the good people at JILA, foremost to my host Stephen R Leone, now Professor of Chemistry and Physics at the University of California in Berkeley, who was instrumental in making my stay possible at this marvelous institution (thanks, Steve). With the inspiration from the sabbatical fresh in my mind, the first incarnation of *It's About Time* got underway. It was followed, after a long hiatus, by three more rounds. With feedback and creative input from the many students in those four instances of the course, the material evolved and somehow morphed into the manuscript *Of Clocks and Time*. I owe thanks to the Wesleyan students, but also to my department who allowed me to teach this course despite the tight curricular constraints we often face. Conversations with and encouragement by my colleagues were a big help in seeing the project through (thanks Brian, Candice, Francis, Fred, Greg, Reinhold, Renee, Tom, Tsampikos). In particular, I recall with gratitude the time and care Edward C Moran from our Astronomy department took to enlighten me about the finer points of cosmology (thanks, Ed). If things are making sense in chapter six of my book, it is because of him (if they don't, it is solely due to my limited grasp of the matter). If there was a tipping point from lecture course to book manuscript, William P Reinhardt, now Emeritus Professor of Chemistry at the University of Washington had something to do with that (thank you Bill). When he gave a special colloquium in my department at Wesleyan, I chatted with him about my course. He was very interested and asked whether I had considered to turn it into a book. I hadn't and I wasn't ready then. But ever since the idea kept popping up in my head—until an email appeared in my inbox soliciting proposals for new book projects with Morgan & Claypool. Then it finally clicked. I want to say thanks to the staff and editors at Morgan & Claypool for their much appreciated encouragement, highly professional support and kind patience. One last and general thanks to the many authors—be it in traditional print or on the web—for writing on time, clocks, calendars in their many contexts. It is a continually growing tapestry that I thoroughly enjoy contemplating.

Author biography

Lutz Hüwel

Lutz Hüwel grew up in a family of seven in Essen in the industrial Ruhr valley region of Germany. He is grateful for the gift of an excellent education and the chance to grow up safe. His parents made that possible, under sometimes difficult circumstances. He is also blessed with four siblings and their families, a set of good friends, and his partner Ruth, his closest travel companion in life's journey.

Schooling included attendance of the modern language branch of the Helmholtz Gymnasium, where the seeds were sown for a lasting interest in all sorts of things and a fondness for reading (a big shout-out to public libraries). Somehow this period also translated into a lifelong fascination and occupation with physics. At the Georg-August University in Göttingen, Germany, he earned the Diplom degree in physics for experimental work on low temperature magnetism. Afterwards, Hüwel was fortunate to be able to earn a PhD at the Max-Planck Institut für Strömungsforschung (MPI for Fluid Dynamics) in Göttingen. Under the guidance of Prof. Dr. Hans Pauly, one of the institute directors, he designed and executed experiments on the scattering of laser excited atoms. Equipped with the additional distinction of an Otto-Hahn Medal for outstanding scientific achievements, Hüwel started a postdoc appointment at JILA in Boulder, Colorado, working on laser investigations of molecular dynamics. A faculty position at Wesleyan University followed. There, in Middletown, Connecticut, he combines teaching physics at the graduate, undergraduate major, and non-science student level with research in molecular photophysics and laser produced plasmas. These studies and their extensions to applied fields such as combustion diagnostics brought about several collaborations with groups in the US, Germany, and the UK.

Chapter 1

Days, months and years

We still do not fully understand exactly how and when Earth and Moon formed in the tumultuous youth of our solar system. A widely accepted theory [1] says that about 4.5 billion years ago, a Mars-sized planetoid collided with Earth and that our Moon is the product of that collision. Presumably, details of Earth's and Moon's orbits are the result of that violent encounter. After things calmed down, still in Earth's early history, stable patterns of motion became dominant. Since then, the ensuing repetitive and nearly constant changes between day and night, between lunar phases, and between seasons have provided powerful natural rhythms that have deeply influenced the evolution of life in general and human development in particular. Obvious manifestations are the myriad biological clocks found in plants and animals alike[1]. Bioregulators may be tuned to 24-hour or circadian cycles, the lunar cycle (for example via the tides), seasonal changes (witness the blooming of asters), or complex combinations (such as the lifecycle of the 17-year cicadas). The natural cadences of time also seem to be reflected in several early manmade structures. The Newgrange (Irish: *Si an Bhru*) mound in County Meath, Ireland, the Stonehenge circles in Wiltshire, UK, or the Kukulkan temple in Chichen Itza on the Yucatan peninsula of Mexico are a few examples. Today we can still experience what people must have felt hundreds and even thousands of years ago when the solstice or equinox Sun aligns with these impressive constructions. While speculations and even fabrications are not uncommon in connection with ancient sites such as these, the influence of astronomical observations on their layout is quite certain. The regularity and reliability with which the Sun marks the same extreme points on the horizon during the course of a year is at the heart of these remarkable edifices. There is also evidence that structures were built in antiquity that aligned with either planets or stars, the other objects that move in regular patterns across the sky [2]. Their influence on earthly events is certainly not of the type that astrology

---

[1] In this book, I will bypass entirely this fascinating and still not fully understood type of clock.

wants us to believe, but the impact of the stars on human thinking, and in particular our thinking about time, is undeniable and strong. One of the oldest artifacts depicting celestial objects is a bronze disc that was found in Nebra, Germany (German: *Sternscheibe von Nebra*), with an estimated age of around 3500 years. On that disc there is, next to the prominent crescent Moon, an arrangement of dots thought to represent the star cluster of the Pleiades in the constellation of Taurus (see figure 1.1).

Well before material artifacts were made that imitated the configuration of the lights in the sky or marked certain aspects of their motion, the rhythms of the seasons—whether directly by their manifestation in the form of dry and wet or warm and cold periods, or indirectly by patterns of animal migration or plant growth—must have left lasting impressions on the human mind. The relentlessness and regularity of that change, even if accompanied by variations and fluctuations, is a powerful indication that something must be 'out there' to keep this rhythm alive. What people across the world and through the ages believed to be the moving agent is a long and fascinating story, but not the one being told here. What we will dwell on is the contemporary version of this story, a version that combines many previous strands into a coherent saga of the Universe. This new tale tells of white dwarfs and red giants, of black voids swallowing everything venturing too close to their edges, and of a mysterious dark substance guiding the ebb and flow of things. It contains chronicles of collapsed stars that spin around their own axis like furious dervishes of old, but with a precision rivaling that of modern atomic clocks. Regarding our closer neighborhood, the saga contains the residue of ancient mythologies letting gods and

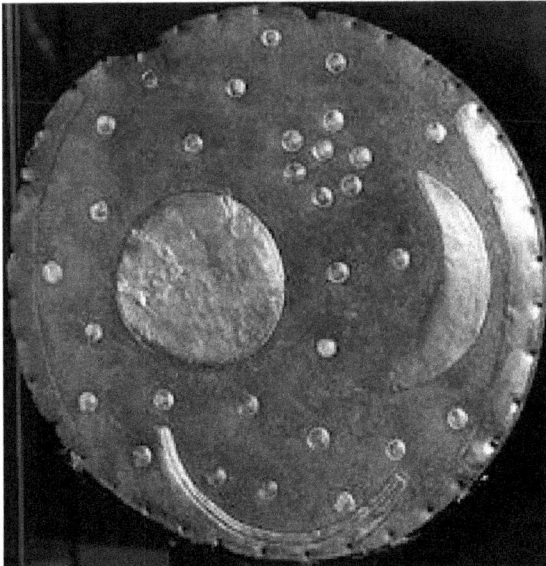

Figure 1.1. Bronze disk, ca 3500 years old, showing waning Moon, Sun and stars—possibly including the Pleiades (between the Sun and Moon). This image has been obtained by the author from the Wikimedia website where it was made available [by name of uploader] under a CC BY-SA 3.0 licence. It is included within this article on that basis. It is attributed to Anagoria.

goddesses of old live on. In their modern version, we can actually see the faces of Jupiter, Saturn, Mars, Venus and Mercury. We have even touched them and have sent messengers to explore their smallest wrinkles, to find out what stuff they are made of and how they became what they are today. In the old stories, the 'dance of the planets' used to be a stately, highly controlled affair confined to our solar system. The new epic reveals that billions of such systems exist whose choreography follows a few basic rules. Nevertheless, each family of planets with its central star—or sometimes two or even more—is different and unique. Our present saga of the Universe is incredibly wide in scope, rich in detail, and exact in its utterances. We can calculate the paths of the planets or artificial satellites with extraordinary precision. We can do that because we have discovered that general and fundamentally simple laws determine the paths and because we have found ways to turn that knowledge into specific predictions of the future. We have come a long way since the Oracle of Delphi.

Gravity is the force that governs motion in the Universe and one of the fundamental laws alluded to above specifies that the strength of gravity falls off with the square of the distance. As a consequence of this law, the path of any two gravitationally interacting objects belongs to the geometric class of conic sections, i.e. it is a curve generated by the intersection of a cone and a plane. Depending on the angle between plane and cone axis, the cut generates a hyperbola, parabola, ellipse or circle. For example, the rotation of planets around their host star occurs in elliptical orbits, often only slightly eccentric, i.e. almost indistinguishable from a circle. Historically, the argument went in the opposite direction: careful observational data (Tycho Brahe) were used to deduce that planets move in elliptical orbits (Johannes Kepler), which in turn led to the discovery of the gravitational force law (Isaac Newton). In any case, all the natural rhythms are consequences of this simple relationship between force law and the geometry of orbits, i.e. between dynamics and shape.

Despite the simplicity of the law governing it, the motion of Sun, Moon and planets also exhibits subtle complexities that for centuries have befuddled both casual and careful observers of the sky. In the end, though, the root cause for this complication turned out to be straightforward and reducible to three major contributions: more than one rotation is involved, the various rotations do not occur in a single plane and the ratios of their respective periods are non-integer. All this gives rise, for example, to seasonal changes of the length of the day, the need for leap years, the difficulty to fit lunar months into years, and a somewhat surprising abundance of definitions of what a day and a year really are. On a finer scale, there are both long- and short-term variations of the basic periods, which have produced further difficulties for anyone trying to come up with a comprehensive model of the motion of the heavenly bodies. The additional variations are caused by the gravitational influence of celestial objects other than the Sun, or the vagaries of the rotation dynamics of Earth's core, to name just a few causes. An example of the former effect is the slow motion of the celestial pole, while the recent practice of inserting leap seconds into our atomic clock-controlled time-keeping is a response to the latter—at least in part. Around the start of the current millennium, a large number of books were published on this very topic—how our calendar, how our clocks, how our sense of time is informed by both natural, objective causes and by social and historical circumstances. While the latter, intriguing aspect is yet another issue that is not pursued in this short book, the former will be. So, how do we describe the celestial clock from our modern perspective? In a moment, we will begin to look into this question. First, we need to introduce some specific terms and units of measurement to characterize the distance between two objects in space, their orientation relative to a given direction and/or plane, and the length of time elapsed between two events.

1.1 A first discourse on measurement, units and precision

At the time of writing, only three countries[2] in the world have not yet officially adopted the metric system, which I will use throughout this book. These decimal-based units are also known as SI units (the acronym stands for the French *Système Internationale*). In particular, we will employ the SI base units of length (meter, unit symbol 'm'), angle (radian, 'rad'), and time (second, 's'). Multiplicative scaling of the base units by factors of 1000 or 1/1000 and attaching associated prefixes (mega-, kilo-, milli-, micro-, etc) provides an expedient way to extend the units to larger and smaller ranges (see also table 3.1). However, in the context of time, additional non-metric units are convenient and commonly used, most notably the units of years, weeks, days, hours and minutes. Likewise, the vast distances encountered in astronomy are usually indicated in non-metric units, such as the astronomical unit, the parsec and the light year.

To measure is to compare with a standard. For example, the distance between two points is determined—at least in principle—by placing an agreed upon stand-ard, say a meter stick, end to end between the two points as many times as is necessary. Clearly, this recipe fails if you want to know the distance to the Moon, the Sun, or any star or galaxy. Whether we are able to reach them directly or not, what can be measured for any two points is their angular separation. Take two stars in the night sky or two points on opposite sides of the Moon's equator. Independent of their distance from us, the two lines of sight connecting an observer with the two points subtend a well-defined angle that can be measured with a simple mechanical instrument such as the sextant. Conversely, two angles specify completely and uniquely the location of any one point in the sky. As with any other definition, we have to choose a reference frame in which the locating angles are to be quantified. One straightforward choice is the local horizon augmented by the four cardinal directions (north, west, south and east). Next, draw an imaginary line, the celestial meridian, in the sky that connects the zenith with any of the four cardinal points on the horizon—say south. Then the first angle measures the vertical angular height of the star above the horizon and the second the horizontal angular separation from the meridian. The first angle ranges from 0° for a point at the horizon to 90° for any object located at the zenith. Stars and anything else on the celestial hemisphere

---

[2] Liberia, Myanmar and the USA.

below the horizon will be associated with negative angles ranging up to −90° for the nadir. The values of the second angle are anywhere between 0° and 360°. For completeness, we also have to specify the direction in which the second angle is being determined along the horizon—say in the east to west direction. This now allows an unambiguous specification of the location of any point in the sky. Of course, the angular values of an actual object depend on the place (and the time) at which the observation is made and observers at different locations will have to convert their specifications. Therefore, astronomers have adopted a 'standard' location from which celestial navigation is conducted. Locally measured angles are routinely transferred to this 'global' reference frame. The discussion of what exactly this reference frame is can wait until we describe the structure and dynamics of our solar system in more detail. Incidentally, position on the surface of the Earth is determined by a pair of angles that are completely analogous to that used to pinpoint locations on the celestial sphere. In the case of terrestrial navigation, the two angles are called latitude and longitude. The former is the angular distance from the equator along the north–south direction (with positive values between 0 and 90° for the northern hemisphere and negative values for the southern hemisphere). Longitude is the angular distance in the east-to-west sense from a reference arc that joins the north and south poles, and which by convention passes through a certain point at the Royal Observatory in Greenwich, UK.

The division of the circle into 360 parts or degrees goes back to the Babylonians and the similarity to the numbers of days in the year is probably not a coincidence. In the SI unit system, an alternative way to express a given angle gives rise to the unit of the *radian*. Since the circumference of a circle of radius r is $2\pi r$, the ratio between the circumference and radius is 2π, which is equivalent to the full angle of 360°. Any smaller angle θ given in degree can be specified by the ratio of the corresponding arc to the radius. The value of this ratio is the magnitude of the angle θ in units of radians. For example, an angle of 60° is the same as 1/6 of the full angle, i.e. $\pi/3$ rad. The metric base unit 1 rad corresponds to an angle of $(360/2\pi)° \approx 57.3°$. Much of the dynamics we encounter in our upcoming discussion is rotational motion—as opposed to movement along straight lines. If the center of the path lies outside the moving body, we speak of an orbit (like that of a satellite about the Earth). If the spin occurs around an axis that passes through the object itself, we refer to that motion as rotation. In either case, a full completion is identical to a change of 360° (or 2π) in angular position of any given point of the body.

Without further ado, set seven days as one week, one day (unit symbol: d) as equal to 24 hours (h), let one hour be exactly 60 minutes (min) long and have each minute contain 60 seconds (s). As already alluded to above, conversion between the time units of year and day is non-trivial and depends on the exact definition of the two quantities. Let us define for now the length of a day as the average time span between two subsequent moments of the Sun being at its highest point in the sky, i.e. from noon to the next noon. Likewise, we adopt the notion that a year is the average time span between two subsequent summer solstices, i.e. moments of the setting Sun crossing the horizon at the most northerly point. Then approximately 365 ¼ days fit into one year. In general, science insists on precise terminology and the preliminary definitions above of the day and the year are two examples. This insistence on precision might be perceived as needless, ('we all know what a day is'), if not outright annoying. But it turns out to be necessary. On closer inspection things are often more subtle or more complex than expected. For example, the two definitions just mentioned include the word 'average', which seems superfluous, yet it is not. As we will see soon, the length of the day and the year is not constant.

With the above definitions, we also have a first set of conversions between time units, which we can write succinctly as follows:

$1 d = 24 h = 24 \cdot 60 \min = 1440 \min = 1440 \cdot 60 s = 86\,400 s$.
$1 a \approx 365.25 d = 8766 h = 31\,557\,600 s$ (note that the precise value is $1a = 31\,556\,925 s$)

In other words, even with a daily 8-hour beauty sleep you can while away almost 60 000 hours in a year or conduct about 2 million back-to-back countdowns—your choice. In order to get a feeling for the implied precision when the specification of the duration of one year is uncertain by one second, it is helpful to consider the precision a clock needs to have to be capable of detecting this last second in the year. Suppose our clock ticks at one-second intervals. Since there are about 30 million seconds in one year, each clock period needs to repeat with a precision that is better than one part in 30 million. While this is no challenge at all for atomic clocks (see chapter 3), only the best mechanical clocks (see chapter 2) can accomplish such a feat.

At present, the basic unit of time is the second, defined as the period of the light wave emitted by cesium (Cs) atoms under certain well-specified conditions. Historically, the base unit of time was the day, defined by astronomical observations, and hours, minutes and seconds were derived quantities. When the atom-based definition of the second was introduced, it was chosen so as to preserve the actual length of the corresponding time spans. This is accomplished with the second fixed as the duration of a bit more than 9 billion (to be precise, 9 192 631 770) oscillations of the cesium atom radiation. In essence, the two sets of equations above are the definitions of the longer units of minute, hour, day and year. It will be useful for our upcoming discussions to imagine that we have available a perfect stopwatch that ticks at a uniform rate and allows us to determine the time span between two events with arbitrary precision. This notion of measuring time spans with clocks appeals to 'common sense', but actually needs refinement, which we will provide in chapter 4 (on relativity). For now, it will suffice.

Adopted worldwide, the Gregorian calendar gives another answer to the question of how many days there are in a year by defining one ('regular') year to be exactly 365 days and adding one day for leap years. Leap years are those whose number is divisible by four, but omitting three out of four century years, namely those that are not divisible by 400 (e.g. the year 1900 was *not* a leap year, but 2000 was). Thus, the Gregorian calendar has a repeat length of 400 years. While certainly not the only one, the Gregorian calendar is a useful and relatively simple tool that comes to terms with the conundrum of incommensurability and lets us organize the flow of time into a pattern of countable units, keeping in sync with the seasons—at least for the next few thousand years. Following the Gregorian calendar, we can calculate the average number of days in one year in the following way. First add up all the days in 400 years if no leap years existed ($400 \cdot 365 = 146\,000$). Then add one day for each leap year, disregarding the century rule ($+100$). Now subtract one day for all centuries not counted as leap year (-3) for a grand total of $146\,097$ days in 400 years. Thus, there are 365.2425 days on average in one Gregorian calendar year. Compare this to the current best value of 365.242 189 days in one tropical year, which is the time span to complete one cycle of the four seasons: the annual difference of 0.000 311 d or about 27 s accumulates to a slip of one day after $1/0.000\,311 \approx 3215$ years. Not bad at all.

Is a calendar, Gregorian or otherwise, a clock? In other words, does it use a 'device' to mark certain basic time intervals, and does it then indicate how many of those units have elapsed between two events? Maybe it is stretching this definition a bit, but I think one can argue that calendars are indeed some sort of clock. The time-beating core is simply the Sun's journey in the sky—which of course means Earth's rotation coupled with its orbit around the Sun. Measuring the elapsed time is accomplished by counting days 'labeled' in a unique manner, for example in the Gregorian calendar by year, month and day, or in the Mayan long count by providing a combination of numbers for the teeth of interlocking gears. Calendar systems have a repeat length, after which the same combination recurs. In this sense, all calendars reflect a cyclic view of time. One long, repeating count in the Mayan calendar is equivalent to $1\,366\,560$ days or about 3741 years, which is similar to that of the Gregorian calendar. For the latter, we have seen that it slips by one day after about 3215 years, which came about because of a mismatch of about 27 s to the average time it takes Earth to complete a trip around the Sun. One might say, then, that the Gregorian calendar, viewed as a clock, runs untrue by about that short time span in one year when compared to the time told by the Sun–Earth system itself. Just to put this into perspective: state of the art atomic clocks perform at such an astonishing level of precision that it takes them millions of years to be off by one second. In chapter 3 we will see how to quantify clock precision with the help of the so-called Allan variance, which, amusingly, also permits a loose comparison of time pieces as diverse as atomic clocks and megalithic monuments such as Stonehenge or Newgrange, which themselves might be viewed as calendar-clocks marking off units of years.

1.2 What we see in the sky—stars, planets, Sun and Moon

Whether via direct or scattered rays, our Sun causes most natural light effects, such as red sunsets, green flashes, blue sky or the whole color spectrum of the rainbow, and the light reflected by our Moon and the planets. There are a few exceptions, including lightning bolts, the occasional meteor or comet and, of course, the fixed stars. Lightning, comets and meteors are light shows that people probably felt more frightened of than entertained by. Although quite frequent, these light effects lack the precision and regularity to contribute much to our story of natural timekeepers—notwithstanding the (near) periodicity with which a number of meteor showers peak in intensity. It is the dance of the Sun itself and of the Moon and the stars, whether fixed or wandering, that had a profound influence on our understanding of the world in general and of time in particular—and our ways to measure it. These lights in the sky have intrigued people for a long time, and not only because of their practical utility. How could it not be so? Imagine that you are living in Egypt, four or five thousand years ago. It is night. Some lingering kitchen fires disturb the darkness, but just a short walk outside the village the sky is pitch dark. The bright splendor of the stars in the night sky must have been mesmerizing—it still is.

Stay awake one night and you are treated to a slow-moving light show (provided you can find a sufficiently dark place): in unison and in an east-to-west direction, the stars complete part of grand circles around a common point in the sky, the celestial pole[3]. Even the fixed stars are not fixed at all. At dawn, they become invisible against the Sunlight scattered by the atmosphere. However, come dusk, the stars reappear, one by one, at the points where a straightforward extrapolation of last night's motion would predict them to be. In other words, the fixed stars rotate through the sky in full circles[4] with a common center and with their positions relative to each other unchanged. This twirl is prosaically called the diurnal stellar motion. During daytime, the Sun completes an arc in the same direction and around the same center. After centuries of fits and starts, we understand that the motion of Sun and stars across the sky is caused by Earth's rotation around its own axis, hence the common center. Earth's spin axis points to the celestial pole. The latitude of any place on Earth determines the declination of the celestial pole. Should you be at the north or south poles, the celestial pole is exactly overhead. Stargazers at the equator can easily infer that there are two celestial poles—one in the north and the other in the south, both near the horizon. If you are more patient and pay attention to the sky all year long, you will further see that the particular star constellation that rises above the horizon in a given direction changes over the course of one year. After that, the same sequence repeats.

Much more restless are the wandering stars or planets (Greek: πλανητησ (wanderer)). Mercury, Venus, Mars, Jupiter and Saturn are visible to the unaided eye and thus they were already known in antiquity. They are not part of any stellar constellation, they participate only approximately in the diurnal stellar motion and their relative motion is not simple. Relative to the fixed stars, they speed up, slow down and sometimes even reverse direction in so-called retrograde motion (see figure 1.2). While Mars, Jupiter and Saturn range freely across the sky—albeit confined to a narrow band around the ecliptic—Venus and Mercury never stray far from the Sun, and sometimes disappear behind it. Probably because of the tight connection to the Sun and its bright appearance as morning and evening star, Venus had a special significance in Mayan culture. Astronomer–priests observed the planet closely. The Dresden codex, one of only four known books written by the Maya, very likely in the Chichen Itza region [3], contains details of Venus' motion in the

---

[3] With a circumpolar path of radius less than 1°, Polaris, the north star, is very close to the celestial pole on the northern hemisphere. No visible star is near the southern celestial pole.

[4] During mid-winter, when the Sun does not rise above the horizon north or south of the polar circles, you can actually observe the full circle of the stars.

Figure 1.2. Observed retrograde motion of Mars (points 2–4) relative to the constellation Cancer in 2009–10. Five sequential observations of Mars' position in the sky are shown by the numbered dots.

sky, including its synodic period of 584 days (the current best value is about 583.92 d). The synodic period of a planet marks its return to the same spot in the sky and ranges from 116 days for Mercury to 780 days for Mars. For the other planets, this interval is much closer to the length of the Earth year, ranging from 399 days for Jupiter to 368 days for Neptune.

For much of human history and across cultures all over the world, the brightest light in the sky—our Sun—has been at the center of both worship and observation. It does not take much sophisticated effort to realize that the Sun's path changes significantly over the course of a year. First, the daily east-to-west arc varies with the seasons in length and height above the horizon. In doing so, the high point of the arc moves up (summer) and down (winter) along a skewed, slender figure of eight called the Sun's *Analemma*, whose slant and shape varies with the observer's latitude. Secondly, the arc changes position relative to the background of the stars. Imagine for a moment (as I alluded to in the preface, we will do a fair amount of imagining) the Earth had lost its atmosphere, which would clearly be a nuisance. Adding insult to injury, there would no longer be any beautiful sunsets or rainbows, and the sky would always be black. The latter aspect, however, would have the advantage of the stars being visible 24 hours a day; the diurnal motion would be in full display even when the Sun was above the horizon. At any given moment, it would be obvious which starry constellation was in closest proximity to the Sun, as we would be able to see the Sun and stars moving along their respective circles in synchrony all day long. That is to say in approximate synchrony, because every day the Sun lags a tiny bit behind the motion of the stars. Relative to the stars, the Sun moves eastward. Thus, the stellar backdrop against which the Sun is seen changes during the course of a year. After one year, though, the cycle repeats. Luckily, we still have oxygen to breathe. Therefore, the moments around sunrise and sunset are fleeting when we can catch a glimpse of a few stars and Sun together. However, since the pattern of the fixed stars is indeed fixed, we can still figure out 'in which house' (as astrology likes to call it) the Sun presently resides. The apparent year-long path of the Sun in the sky against the background of stars defines a plane in the sky, called the *ecliptic*, whose axis is tilted by about 23.5° against the axis of the diurnal motion. During our ride on the merry-go-round we call the solar system, when we look towards the center where the Sun is, the far distant background scene of stars is constantly changing as we complete one circle.

If you are very patient and pay attention to small details, you can also see something very subtle—the celestial pole moves relative to the stars. It shifts ever so slightly—and with it the point on the horizon where stars rise or set that are sufficiently far from the pole. Earth's spin axis describes a circular path against the fixed stars and it takes, in round numbers, 26 000 years for one completion. This phenomenon—equivalent to that of a wobbling top—is primarily a result of the gravitational tug by the Sun and Moon on the Earth's bulge around the equator. As a result, both the north and south celestial poles wander by about 1.4° every century. While not all that much, it still amounts to almost three times the angular size of the Moon. Astronomers refer to the long-term celestial pole motion as 'axial precession' or 'precession of the equinoxes', a term introduced because the shift is observable as the westward movement of the spring and autumn equinoxes along the ecliptic relative to the fixed stars. Because of Earth's pole motion, which stars if any are closest to the celestial poles changes over time. Currently, we find the star Polaris (aka α-Ursae Minoris) closest to the North celestial pole. When the last glacial period was still in full swing about 14 000 years ago and you looked up in the sky to find your way home at night, the brightest star near the pole would have been Vega in the constellation Lyra [4].

And then there is the Moon. Its angular size, quite accidentally, is very nearly the same as that of the Sun. As beautiful as the stars and the Milky Way are, our Moon outshines them all, albeit with secondhand light. Waxing, waning and distinct features in the face of the Moon add to the allure. No wonder tales about the Moon and, importantly in our context, attempts to build a lunar calendar can be found in many cultures. The time span between two full Moons (or any other lunar phase) is called the synodic month. On average, it lasts roughly 29.5 days but varies by as much as 18 hours during the year. Since about 12 ½ lunar cycles occur in one solar year, lunar-based calendars are very difficult to synchronize with the solar rhythm.

The Sun and Moon follow independent paths and the Moon's arc across the sky is slanted relative to the ecliptic, thus allowing for crossings. Occasionally the Moon indeed moves in front of the Sun, obscuring it partially or even fully. On the other side of its orbit, the Moon can dive into the shadow cast by the Earth, causing a lunar eclipse. Solar and lunar eclipses occur not infrequently but they lack simple periodicity. Nevertheless, attempts to predict, for example, total solar eclipses started surprisingly early, even if initially they were not blessed with much success. As early as 2300 BCE, the Chinese emperor apparently expected solar eclipses to be foretold. At least that is how it appears, since two astronomers at the imperial court were beheaded for failing to predict the eclipse of that year [5]. Despite this strong disincentive to pursue the issue, by the 4th century BCE Chinese astronomers had learned to predict solar eclipses by observations of the relative position of the Sun and Moon. As we will see later in the book, in modern times a solar eclipse has played an important role in our advancement of the understanding of time. Yes, you guessed it, I am referring to the solar eclipse of 1918 confirming Einstein's prediction of light bending by gravity. In chapter 4, we will discuss another wrinkle concerning the Moon's role in the context of general relativity.

Associated with the motion of the Sun along the ecliptic is a very small, but nowadays easily measurable, variation of the angular size of the solar disc. In January, the Sun's angular diameter is found to be slightly more than 3% larger than it is in July. Let us discard the highly unlikely hypothesis that the Sun cyclically changes its actual size in response to or at least in synchrony with the seasons on Earth. Then the periodic change in angular size hints strongly at a small change in the distance between the Sun and the Earth—a little closer than average during the winter in the northern hemisphere, a little further away in the summer. By the way, appearances notwithstanding, there is *no* measurable change in the angular size of either the Sun or the Moon during the course of one day. They do *not* increase with approach to the horizon, although refraction in the atmosphere does distort the image.

Let us now return to the question of how the natural units of the day, the year, or the synodic month arise by looking at the make-up of the solar system, the ultimate mechanical clock. It may appear as a useful and probably necessary detour to describe the spatial properties first in order to discuss the temporal ones—but it is a detour nonetheless. However, this seemingly indirect route already hints at a deep connection between time and space, as fully revealed only in Einstein's relativity theories (see chapter 4).

1.3 What the solar system looks like

Once upon a time, we thought that we had figured it out. Earth is at the center of the Universe with the Sun, Moon, a few planets (the wandering stars) and the fixed stars circling around. The various celestial light sources are attached to 'crystalline' spheres[5] that are nestled like spherical Russian dolls and that glide smoothly around each other in simple harmonic motion at their own unchanging speed. The Moon is on the sphere closest to Earth, and then come Mercury, Venus and the Sun, followed by Mars, Jupiter and Saturn, and finally there is the firmament with all the fixed stars. This, in a nutshell, is the cosmological model that was developed in ancient Greece and survived in one form or another for more than a thousand years. The model is efficient, as it reproduces all of the salient observations with just a few explanatory ingredients. In addition, it is flexible enough to accommodate, with some tinkering, additional details such as the retrograde motion of the planets. Single spheres, rotating around the Earth with constant angular speed, just cannot explain it. However, if the planets are allowed to move in smaller spheres attached to

---

[5] To modern ears it sounds strange and naïve to have whole planets suspended by such a fragile substance as quartz. But that is in hindsight—the early view of the 'lights in the sky' was just that: lights in the sky. One can argue—I think convincingly (see e.g. [6])—that the geocentric model is an early scientific attempt to come to terms with observation, 'explaining' nature without recourse to supernatural powers.

the large sphere that carries the planet (so-called epicycles), then the observed motion can be reproduced. Adding more and more epicycles, at the cost of its original simplicity and elegance, the evolving geocentric model was able to duplicate all relevant observations as late as the 16th century. In addition, the model has the benefit of being consistent with the immediate impression of our senses, telling us that we are standing still and that the Sun, Moon and stars are moving around us. It was a truly revolutionary and at first fiercely resisted insight that the observed motion of the Sun is only apparent and that it is the Earth that swings around a (nearly) stationary Sun.

Here is what we now know about our solar system and the stars. Not Earth but the Sun is at the center of the dance. Earth is but one of a huge number of objects orbiting Sol, our central star. The Sun dominates the total mass of the system: the estimated contribution to the total mass by all other objects is only about 0.14%, in other words the solar mass is about 700 times larger than the mass of everything else in the solar system combined. We can classify these other objects according to size (or mass) and orbital shape into categories (asteroids, comets, planets, etc) whose boundaries are not entirely fixed—witness the demotion of Pluto. Planets are the objects found in the inner part of the solar system that are massive enough to be round due to their own gravitational force as well as to be able to sweep clear the vicinity of their orbit. They move, all in the same directions, around the Sun in near circles with radii that differ by a factor of almost 80. In terms of the Earth–Sun distance, which we introduce here as one astronomical unit or 1 au, the distance of the innermost planet Mercury is about 0.39 au, while that for Neptune, the outermost planet, is about 20 au. In the context of this book, the planets and their satellites are our immediate concern. Of the former, the solar system contains eight, with Earth being the third rock from the Sun. It started with Galileo (see figure 1.3) and by now, astronomical observations have discovered 178 moons to orbit six of the planets. Venus and Mercury do not have known companions, and Earth and Mars have one and two, respectively. All planets

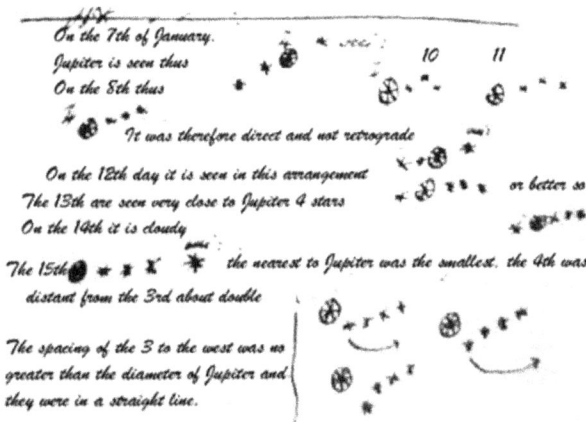

Figure 1.3. Page from Galileo's manuscript detailing the day-to-day alignment changes of Jupiter's four largest moons. Source: Regents of the University of Michigan (CC-BY-SA-4.0).

revolve around the Sun in non-crossing elliptical orbits, with the Sun in one focal point. The orbital planes of the planets are somewhat tilted against each other. Planetary satellites are also moving in planes and these are tilted against the orbital plane of their respective hosts. Finally, the planets turn around an axis through their own center. These axes are tilted yet again relative to the planet's orbit. However, this widespread tendency to be askew is, for the most part, only a matter of degree and does not lead to random orientation. Deviations from the mean of the planetary orbit tilt in particular are not very large. The corresponding axes of the planets Mercury to Uranus vary around the average by only about ±3.5°. In other words, all planets move in the same sense in a disc around the Sun. Most planets rotate around their inner axis in the same sense as they orbit the Sun. However, there are two odd balls—Venus is spinning retrograde and Uranus' polar axis lies almost inside the orbital plane. The next section will discuss in more detail how the various orientations and alignments can dramatically affect how time goes by as measured by the course of the Sun in the sky.

All time-keeping derived from celestial observations is ultimately rooted in rotational motion, be it via orbits around an external center, namely the Sun for the planets or the host planet in the case of satellites, or via intrinsic rotation around their innermost core. Rotation is repetitive and consistent with a view of time as being cyclical. A linear and directed quality of time—which has been dubbed the 'arrow' of time—arises from the uniqueness of history and thus from the breakdown of complete periodicity. At first and even second glance, celestial dynamics is cyclic. However, when you apply high precision to short-term studies or exercise patience and consistency in conducting long-term observations, you will discern deviations from purely repetitive motion. The former approach is a modern one, while Babylonian and Egyptian early stargazers were already applying the latter thousands of years ago.

1.4 What a day makes

We have seen that Earth orbits the Sun in one year and rotates around its own axis in one day, and that the Moon orbits Earth in about 29.5 days. These terse statements define three separate units of time. When we say that something has completed 'one rotation' or 'one orbit', we mean that the angle measuring the position of the corresponding object or feature has changed by 2π radian or 360°. Although the Sun and stars track similar circles in the sky, we specifically use the Sun's motion to define the length of one day. Returning to the same spot in the sky marks an average time span of 24 hours or 86 400 seconds. One might expect that substituting a star, any star, for the Sun and measuring the star's completion of one round in the sky would delineate exactly the same length of time. Let us take a good clock and measure. What we actually find is that the stars complete their daily circles faster than the Sun does—by about 235.9 s, a little less than four minutes. The geocentric model built on celestial spheres recognized that fact by letting the stellar sphere move a bit swifter than the sphere of the Sun. Originally, this adjustment was not implemented on account of the small difference per day, but rather because of the annual shift of the Sun relative to the constellations of the Zodiac. In any case, we now have two different definitions of the time unit 'day'—the solar day of 24 hours and the sidereal day of about 23.934 hours. Multiplying the difference between solar and sidereal days by the number of days in one year yields the annually accumulated difference. Interestingly, we find this difference to be equal to the length of one additional sidereal day (365.2425×235.9 s $= 86\,161$ s $= 23.934$ h). In other words, during the course of one year, i.e. one complete orbit around the Sun, there is one more revolution of the Earth around its axis relative to the stars than there are rotations as measured relative to the Sun. This is no coincidence.

Thanks to a string of discoveries made around the beginning of the 21st century, and starting as early as 1988 with the first confirmed sighting, we know that at least in our own galaxy planets are abundant. It appears that planets might even be a necessary by-product of star formation. Imagine that we have perfected our instruments and that we can not only measure the motion of planets around their central star with great precision, but also the spinning around their own axes, as well as any geometrical aspect of their motion, such the eccentricity of the orbit and the tilt of the rotation axis relative to the orbital plane. In order to illustrate the difficulties faced when deriving a calendar from the observed changes of the Sun and the Moon, let us consider a few idealized cases. We have not (yet) actually observed these cases, but none would violate any known law of physics. Ideal systems just make it easier to reveal the fundamental issues. Who knows, maybe there are some civilizations out there that developed their calendars and clocks in such environments.

Suppose that one of the exoplanets orbits a star (let us call them Aplanet and Asun) in a mathematically perfect circular path. Unlike Earth, Aplanet's spin axis is oriented exactly perpendicular to the plane of orbit around Asun. Also unlike Earth, the exoplanet does not have a satellite. For observers such as us looking at the system from afar two obvious time scales exist: the time for Aplanet to complete one circle around Asun (one Aplanet year or one Ayear for short) and the time for Aplanet to finish one rotation around its own axis, a period we will call one Aday. The magnitudes of the two time scales are completely unrelated to each other because they are determined by accidental details of the formation of this star–planet system. Furthermore, the orbital and spin rotation could be in either the same or the opposite direction. Whatever the ratio, Aplanet cannot sit still in its place relative to Asun. Gravity would pull it very quickly towards the center and poor Aplanet would be gone in a flash—literally. The possibility exists, though, that Aplanet does not rotate at all around its own axis. In that case, would there be night and day on Aplanet? Suppose an Aplanetian living at the equator of Aplanet firmly plants a stick vertically into the ground. Because there is no rotation, the stick will always point into the same direction in space—by accident towards us[6]. From the stated assumptions it follows that we happen to be in the plane defined by the orbit of the

---

[6] There is a slight cheat here. I am describing the situation as if the geocentric model were correct. However, the distances to exoplanetary systems are so large that it makes no significant difference whether the stick points to us or to the Sun. To be entirely in the clear we can position our viewing platform somewhere on the line between Asun and our Sun.

distant planet. Therefore, at certain moments Aplanet will be exactly on our line of sight and in front of Asun and half an Ayear later exactly behind it. While Aplanet is in front of Asun from our point of view, the stick points away from Asun. Thus, the area surrounding the marker is dark. When the Aplanet is behind Asun, the stick— still pointing towards us—now also points towards Asun and the area is lit. Our ingenious Aplanetian will measure the length of day and night to be exactly equal to the period we measure for one orbit. A different question is how she could ever know for sure that Aplanet is in orbit. Other than the Asun's slow ascent and descent to and from the zenith, nothing is changing from one day (or year) to another. Thus, by the way, there is no clear distinction on Aplanet between annual seasons and daily weather, which presumably does exist, driven by the varying irradiation of the atmosphere during Asun's protracted rise and fall in the sky.

At the exact moment when Asun is at the zenith, the vertical stick will not cast a shadow. After that, Asun will move along a great circle in the sky, with the stick casting an ever-longer shadow. Once Asun touches the horizon, the light dims and, after some twilight, the fixed stars appear in the dark sky. Because Aplanet is the lonely companion of Asun there are no 'wandering stars', and no other planets to look at. What will the fixed stars look like? Remember that our assumption is that of no rotation relative to us, the distant observers who are in turn essentially stationary in relation to the fixed stars. Looking up overhead along the direction of the stick, a certain star might be visible. To aim the stick at another star, you would have to tilt the stick somewhat. No matter which star you choose, the stick will always point at the chosen star and will not wander away. On Aplanet, there is no diurnal stellar motion. Of course, the stars grow dimmer and eventually become invisible as dawn approaches and the sky grows lighter (Asun appears from below the horizon exactly opposite from where it sank below the horizon half an Ayear ago). If the Aplanetians have evolved on a planet without atmosphere, the stars remain visible. So here is an interesting finding: the central star of Aplanet is seen as 'moving' in the sky while the stars do not move at all. They are truly 'fixed stars'—at least over times scales that are short compared to any large-scale motion of this mini solar system. A would-be Acopernicus would face a daunting task to discover what we can see clearly, namely that Aplanet orbits Asun. But even in this instance there is a way to observe orbital motion—see the discussion of stellar aberration in section 4.2.

In terms of sidereal days and solar days, we can succinctly describe the above scenario by stating that, in the absence of planetary rotation, for each orbit there is one solar day and zero sidereal days (one might also say that the sidereal day never ends and thus is infinitely long). Could there be the reverse situation, so that there is an everlasting solar day and one sidereal day per orbit? The answer is 'yes'. Instead of discussing another hypothetical exoplanet (in which case I would be tempted to go backwards in the alphabet and name them Zsun and Zplanet), we will train our sight on a much nearer spectacle to illustrate what can happen when orbiting and rotation are combined. Meet Mr P!

Occasionally, Mr P enjoys a somewhat unusual ritual in the local park where he walks around a statue in a precisely choreographed manner to celebrate the cosmic dance of planets. Here is how he does it.

Beginning simple, the first act is dedicated to planets whose solar days and years are indistinguishable, just like the Aplanet we just visited. Mr P starts at a certain distance away from the statue, looking straight at it and simultaneously at a tall building, located much further away. From here on, Mr P fixes his gaze solely on the distant building, never to lose sight of it or change the direction in which he looks at it while he circumnavigates the statue. In this scenario, the statue will obviously disappear behind Mr P's back after half of the circle. In this analogy, the Sun (aka statue) shines during half of the planet's orbit (aka Mr P's circle); during the other half, the Sun is below the horizon and it is night. In other words, one solar day is exactly equal to one year and the sidereal day is infinitely long, since the stars never change position in the sky (or the building is always located in the same direction). This choreography is essentially the scenario just described for the Aplanet–Asun system.

Act two shows how a sun can stay permanently still in the sky. Throughout his circular path around the statue, Mr P now steps sideways and thus always looks at the statue. In doing so, the distant building gradually moves out of his sight (of course, he and not the building is doing the moving). After half a circle, the house has disappeared behind his back, only to reappear into his exact central vision after one full circle. Does Mr P turn around his own axis? Anyone watching from the building will certainly think so. From that vantage point, one sees Mr P's front at one point, then one of his sides, his back, the other side, and, upon returning to the initial spot, once more his front. So, yes, relative to the 'background' defined by the building, he does rotate around his axis. From Mr P's perspective, the statue does not move—the Sun remains stationary in the sky all the time. However, the building and all other points at the horizon move in a circular fashion—the stars are moving in grand circles while keeping their relative distances and relative positions constant.

Specifically in the case considered, orbiting and rotation take place in the same direction, say both clockwise when viewed from above. What would happen if they occur in opposite directions, but with the same period? Assume Mr P faces the statue and building when he starts his dance. After half a period, he has completed half a rotation around his axis and thus now looks away from the building. He has also finished half an orbit and thus looks again towards the statue. In between, after one quarter of a period, he will face away from the statue. During each round-trip, he will face twice towards and twice away from the statue, but only once towards and away from the building. In other words, there are two solar days but only one sidereal day per sidereal year. We do not have to travel to distant systems to find a planet like this: Venus in our solar system is very nearly performing such a dance (see figure 1.4). As already mentioned, Venus does rotate in a retrograde direction, and does so very slowly. Even today, the reasons for this odd behavior are unknown.

We have seen that the mere act of orbiting without any intrinsic rotation already yields one solar and zero sidereal days per revolution. In the absence of orbiting, the number of sidereal and solar days is equal. Therefore, in combination the two numbers differ by one. Depending on the relative sense of rotation and orbit, the 'extra' solar day from the orbital motion is either added to or subtracted from the number of days arising from the rotation around the planet's axis. How many days

Figure 1.4. The orbital position and rotation of the planet Venus shown at 10 Earth-day intervals from 0 to 250 days. The position of the point of the surface that was the antisolar point at day zero is indicated by a cross. Reproduced from [7].

fit into the time span of one orbit obviously depends on the relative magnitude of the two associated periods. Because of gravitational influence and the internal friction of planets and their moons, the various periods do change slowly over time. In the Earth–Moon system, this interaction has led to a strict 1:1 synchrony of the Moon's rotational and orbital periods—Mr P's second dance.

In either of the two illustrations, a clear reference frame exists—the distant building, the distant stars—against which the definition of motion becomes clear and unique. In particular, the rotation of the planet around its own axis is evident and well defined. It is crucial that such a reference frame exists, a condition that turns out to be far from trivial. Our galaxy, for example, rotates around its center, carrying the solar system and us with it in a gigantic circle that completes in one *galactic year* (about 225 to 250 million years). Therefore, our vista of extragalactic objects and their alignment with the stars of our galaxy changes dramatically in one galactic year[7]. On much shorter time scales, proper, relative motion of stars in our galaxy is nowadays easily detectable. Once we accept that the stellar diurnal motion is only an apparent motion caused by Earth's rotation, the fixed stars are a good first choice for a space-fixed reference frame. It will do fine as long as we are not concerned with finer details or large-scale structures and long-term dynamics.

---

[7] Interestingly, much of cosmic history can be conveniently expressed with this time span. In galactic year units the age of the cosmos is about 61, that of our sun about 18, and the Cambrian explosion of life forms happened less than three galactic years ago. As useful as the SI units are, systems or processes can often provide 'natural' units that are more commensurate with the prevailing scale.

As we will see soon, Newton picked exactly this reference frame to represent an assumed 'absolute' space.

1.5 Discovering the laws of motion

In order to be consistent with the available data the geocentric model underwent significant changes, in particular the addition of nestled off-center circles to accommodate the complex motion of the planets. However, in the effort to stay abreast with observation, the model had become very cumbersome. Copernicus did *not* improve the quantitative agreement between theory and observation when he asserted—but initially did not dare publish—that the motionless Sun is at the center of the Universe or that while the Moon does revolve around the Earth, Earth itself and the other five known planets orbit the Sun. He certainly did not help the matter by asserting that the planetary orbits are circles with harmonic proportions of their radii. In fact, because Copernicus continued to insist on the circle as the foundation for all celestial motion, he encountered the same problems as the geocentric model when it came to incorporating the known and complex motion of the planets. Nonetheless, his work was radical and Copernicus agreed only after extensive prompting and prodding to publish his theory in the book *De Revolutionibus Orbium Coelestium*. When this indeed revolutionary manuscript eventually came out in print in 1543, it did not contain any discussion about what might *cause* the motion of the planets or what could sustain their orbits. In that sense, Copernicus was still quite close to the Ptolemaic worldview of eternal spheres that simply exist and need no further explanation.

Galileo was the first to see them—four tiny specks next to Jupiter (see figure 1.3 once more). One moment they lined up this way, a few days later a different way. Then they disappeared, only to come back into sight somewhat later. The change repeated like clockwork. With the help of the recently invented telescope, Galileo had discovered Jupiter's largest moons. He also found craters on our own Moon, stars that are invisible to the unaided eye, sunspots and Saturn's rings. The booklet *Sidereus Nuncius* ('Starry Messenger'), published in 1610, summarizes these remarkable observations. They shook Galileo's faith in the geocentric worldview and led him to support publicly the Sun-centered cosmology of Copernicus, a stance that famously got him into quite a bit of trouble. There is much more to Galileo's work than otherworldly discoveries. His astute examination of everyday objects and imaginative experimentations led him to insights that are part of the foundation of modern science—in particular and most relevant in the current context, the science of motion. Up until then the Aristotelian view held sway, in which two kinds of motion exist: one for ordinary matter, whose properties rest entirely on four elements, and another for the heavenly crystalline spheres, which rotate eternally without change and are made of a fifth, ethereal element (aptly called quintessence). Specifically, natural things released from the same height fall faster when they are heavier. Aristotle is silent when it comes to the behavior of objects *during* free fall. To be fair, this was difficult to observe, given the tools available at the time. Today we can use movies of the process played back in slow motion. Galileo did something equivalent—he slowed down the motion itself with the help of wooden ramps with variable tilt and smooth, straight grooves along which a polished brass sphere could roll downward. When the tilt angle is small, the sphere moves very slowly, allowing us to measure details of its motion. Galileo's arguments took off from the change in behavior with increasing tilt angle, extrapolating to free fall as the asymptotic limit of very steep incline angle. Incidentally, as a stopwatch Galileo used water flowing out of a vessel via a thin pipe into a measuring cup. The weight of water accumulated in the cup, determined with a balance, is then a measure of the elapsed time. From his careful measurements, Galileo found the correct relation between time and speed, namely that the speed of objects in free fall increases in proportion to the time spent since their release from rest. He also surmised that all objects fall at the same rate once drag and friction have been removed.

The final breakthrough to a modern science of mechanics had to wait for Isaac Newton, who was born in 1643, within a year of Galileo's death. It also had to wait for another level of precision and completeness in the astronomical database, as well as a refinement of the Copernican model. Between 1686 and 1687 Newton published his theory of dynamics (and so much more) in the three-volume *Philosophiae Naturalis Principia Mathematica*. In the preceding decades, others—some of the giants on whose shoulders Newton famously proclaimed to be standing—had accumulated a large body of ideas and reliable observational data on the motion of various objects. As already mentioned, Galileo in particular had measured the details of objects falling to the ground on Earth, including projectile motion. The Danish astronomer Tycho Brahe (1546–1601) is responsible for a similarly accurate and precise body of data related to objects in permanent free-fall around the Sun (aka the planets) or Earth (aka our Moon). Brahe was a colorful man with a rather colorful biography that includes losing part of his nose in a duel. He was also fabulously wealthy. On top of that, he managed to coax the king of Denmark and other patrons into giving money for an array of buildings and specially designed star gazing tools. The result was Uraniborg, at the time the most sophisticated center for astronomical observations, at least in Europe. Although no telescope was at their disposal, Brahe and his assistants were able to measure a vast number of ascension and declination angles of the planets with unprecedented precision. Simultaneously recording the time at which the planet's position in the sky was observed allowed the eventual reconstruction of the planet's motion from the data. In particular, extensive and precise data for the motion of Mars became available. The high quality of the observational data attracted a lot of attention, including that of Johannes Kepler, who in 1600 joined Brahe as an assistant in Prague. Brahe's new position as the imperial astronomer of the Bohemian king had a lot to do with deteriorating relations with his own monarch at home (as I said, Brahe lived an interesting life). Kepler's motivation was quite simple by contrast: he wanted access to Brahe's records. Of course, the senior astronomer did not just hand over his studies, but rather offered the junior partner a position in his firm. As one of the first tasks, Brahe asked Kepler to try to solve the retrograde puzzle of Mars using the detailed observational records. The challenge turned out to be more difficult than either one of the two had probably thought. It took Kepler more than a decade to solve the puzzle, but he did solve it. It turned out that observed facts, in particular Mars' apparent backward motion, were much more readily explained on the basis of all planets moving along *elliptical* orbits, rather than the harmonic circular motion invoked from antiquity up to Kepler's time, most famously by Copernicus. With that simple change, Copernicus' heliocentric model became a powerful tool to describe even minute details of the witnessed motion of the wandering stars. Kepler was able to summarize his findings into three rules (the first two in 1609, the third in 1618).

(1) Planets move along paths that have the shape of an ellipse, with the Sun in one of the focal points.
(2) An imaginary line connecting a planet and the Sun sweeps out equal areas in equal times, regardless of the planet's location on the ellipse.
(3) The squared time for completing one orbit—as judged relative to the fixed stars—is in proportion to the cubed major axis of the ellipse.

The clockwork in the sky begins to surrender its mystery, permitting much clearer views of some of its inner workings. The celestial clock has several hands, all moving in elliptical paths around the Sun as the central pivot. Mercury completes approximately 120 solar round trips for each Saturn orbit. So you can think of Mercury as being a half-minute hand, while Saturn shows the hour. Knowing the ratio of the periods of a given pair of planets for completing one full round around the Sun, you can compute from the third rule the ratio of their respective distances from the Sun—or vice versa. According to the second rule, planets move at unequal speeds along their elliptical orbits. Specifically, they move fastest when they are closest to the Sun and slowest when furthest away, i.e. near the 'empty' second focal point. That this is so follows from the fact that the connecting line to the Sun is shortest in the former instance and longest in the latter. Thus, in order to sweep out equal areas, the planet's speed has to adjust accordingly and as stated above. These examples show that by measuring time, spatial aspects of the solar system become apparent and geometric portions reveal temporal relations. A linkage between time and space already exists in classical physics. How this connection becomes even tighter in Einstein's twin theories of relativity is the topic of chapter four. For now, let us get back to our immediate story.

Typically, answers in physics—or in science in general—lead to more questions. How come the planets are so good at knowing Kepler's rules? I mean, they did not take driving lessons and just learned how to behave in traffic. A better question would be, what drives them to behave this way? During Newton's time, the general and somewhat vague notion had taken hold that gravity choreographed the dance of the planets. Several people (Newton himself names Wren, Hooke and Halley) had suggested that the gravitational force diminishes with increasing separation, specifically as the inverse square of the center-to-center distance. But this was based more on guessing than on quantitative arguments. At the same time, Kepler's proposition of elliptical orbits gained more widespread acceptance and some even tried, without success, to join the two ideas. So when they met at Cambridge in 1684, Edmond Halley (yes, the comet is named after *him*) asked Newton what he imagined the shape of orbits would be if the planets were attracted by the Sun with a force that weakens as the inverse square distance? Then not-yet Sir Isaac replied without much hesitation that he had mathematically proven that those orbits would have to be elliptical. At that moment, he could not quite find the pertinent notes, but not to worry. In three months' time Newton sent a paper (*De motu corporum in gyrum* or *On the motion of bodies in orbit*) mathematically proving that Kepler's elliptical orbits are indeed a consequence of the gravitational force law. After that, the floodgates opened. Within two years, Newton, with continued prompting and financial support from Halley, had published the three volumes of his masterpiece, the *Principia*, which is undoubtedly one of the most influential science books ever written. For our purposes, we are interested in the three laws of motion, which we express in the following, paraphrased form, with their modern symbolic expression given in parentheses.

First law   Unless a non-zero force acts on them, objects remain at rest or in uniform motion along straight lines (v = constant if $F = 0$).

Second law When a non-zero force acts on an object, the velocity of the object changes in proportion to the strength and in the direction of the force and in inverse proportion to the mass of the object ($dv/dt = F/m$)

Third law   To every action there is a reaction of equal strength and opposite direction. This is in particular applicable to two objects interacting with each other ($F_{1\text{-on-}2} = -F_{2\text{-on-}1}$).

As self-evident, limited, or boring these statements might appear to some, they express deep and important insights. Much of modern engineering and manufacturing owes its efficiency, if not its entire existence, to these equations. More importantly for the context of this book, they also incorporate assumptions about space and time, which will be the topic of the next section.

1.6 Absolute space and time

One of the benefits of Newton's equations of motion is their practical utility. I tell you the position and velocity of an object at one moment in time. In addition, I specify the force acting on the object in a certain region of space. With that information, Newton's second law enables you to predict the position and velocity of the particle at *any* other time, as long as the object does not leave the region. You can predict the future—for example, the date and time of the next lunar eclipse. The more precisely and accurately you can identify the force and initial conditions, the more precisely and accurately do you know the entire path of the object. Space probes are another example. Whether they journey 'only' straight to the Moon or more ambitiously to Pluto on an intricate trajectory involving several gravitational slingshot maneuvers, the flight plan to get them from A to B is based on the procedure outlined above. The Jet Propulsion Laboratory in Pasadena in California is home to an astonishing database and set of equations, the Ephemerides [8], that model the dynamics of our solar system—planets, satellites of planets, asteroids and comets. At its core are Newton's equations of motion and an ever-growing set of positional observations. Think of this model as a digital orrery, a virtual painting of the detailed comings and goings of Mercury and Mars, Venus and Neptune, Titan and Europa—the whole pantheon of Greek mythology resurrected in a computer.

In addition to their inherent utility—certainly not only for astronomers—the laws of motion contain basic ideas about space and time that are still relevant today. The first law stipulates that in the absence of a force, objects either stay at rest or move with an unchanged velocity. Is that reasonable? You managed to get a window seat on your next flight to Paris (lucky you) and the tarmac is rushing with increasing speed past you as the jet is about to take off. If Newton's laws are valid, then you are compelled to say that a force acts on the tarmac, heck on the entire ground plus airport and all. Of course, you well know that no such thing takes place. While Newton had no inkling about airplanes, he was well aware of the existence of accelerated frames of reference such as the one just described. The wording of the first law indicates that his theory only applies to reference frames in which acceleration of objects comes about because of an actual force. These systems also go by the name *inertial reference frame*[8]. The interior of an airplane taking off is *not* an inertial reference frame. Newton understood the motion of inertial reference frames as occurring in absolute time and relative to absolute space, and therefore a special reference frame exists that is at rest relative to space. Space is the container in which all objects exist, the stage on which all events unfold. A universal clock against whose rhythm the pace of every change is measured tells absolute time. Both space and time exist apart from and independent of any material object. Not everyone bought into these concepts, including Gottfried Wilhelm Leibniz (1646–1716), a polymath and philosopher who took issue with the idea of space and time existing in isolation. Despite such opposition, the concept of absolute space and time took hold of physics for the next 200 years and more until the beginning of the 20th century, when a patent clerk ... well, I am getting ahead of myself. That part of the story has to wait.

Newton's theory of motion has more to say about time: it offers a new approach to the notion of the 'now', the instantaneous moment of (or in?) time that is analogous to a mathematical point in (or of?) space. Newton fashions this contribution using a new tool, the mathematics of differential calculus, which he invented for the explicit purpose of applying his theory to concrete real-life situations. In the above summary of the three laws, I included the modern mathematical expressions that express the same content in a highly condensed way—in particular, the differential calculus expression dv/dt for the rate of change of the velocity, i.e. acceleration. Of course, *rate* was and is a familiar concept even without calculus. We all know—or at least should know—that our bank account will deplete if the spending rate is larger than the rate at which money flows into the account. The level in a leaky bucket will rise if you pour in water faster than it flows out. As long as the rate of change is constant, we can also easily calculate the overall change. Suppose you accelerate your car from rest at a rate of 10 m s^{-2} (or a in general). After 10 seconds (or t in general), your car will run at a speed of 100 m s^{-1},

---

[8] In chapter 4, we will explore the significance of inertial and accelerated reference frames in the context of Einstein's theories of special and general relativity. Also, for a good explanation why the first law is not just a redundant consequence of the second, see, for example, [9].

about 100 km h^{-1} (or $a \cdot t$ in general). Easy, but not realistic. If the acceleration were truly constant, the speed of the car would grow above all bounds, which clearly does not happen—for better or for worse. In reality, due to air drag and the limitations of the motor, the acceleration will diminish with time and eventually the car will reach its top speed. Calculations become more challenging when rates are not constant and that is precisely why and where we need differential calculus. If the acceleration of our car varies, how can we figure out how fast the car runs at any given moment and how far it has traveled since the start at time $t = 0$? If you are willing to dispense with accuracy, you can use a simple recipe to find approximate positions and velocities at discrete moments of time t_1, t_2, t_3 separated by a short time interval Δt (i.e. $t_2 = t_1 + \Delta t$, $t_3 = t_2 + \Delta t = t_1 + 2\Delta t$, etc). Just as abbreviations in cooking recipes (tbsp = tablespoon, tsp = teaspoon, °C = degree Celsius, etc) are useful and common, we make use of writing a_n for the value of the acceleration at time t_n, with $n = 1, 2, 3 \dots$ enumerating the time steps. In order to simplify matters further, let us assume that the motion is along a straight line (we hold on to the steering wheel and hope for the best). With that, the recipe reads as follows.

1. As the king told the white rabbit, begin at the beginning—where you know position and velocity. For our example of the car, we have $x_1 = 0$ and $v_1 = 0$, respectively. Because force and mass are specified everywhere, you also know the acceleration at that moment.

2. From the data in step 1, calculate the velocity at the next moment t_2 as the sum of the previous velocity and the velocity change due to the acceleration acting during the intervening time interval, i.e. $v_2 = v_1 + a_1 \cdot \Delta t$. Note that this assertion is not exactly correct, because the acceleration is *not* constant. Therefore, a better choice for the acceleration would be the average acceleration between times t_1 and t_2. However, if the time interval is short, the error will be small. Because the car is initially at rest, i.e. $v_1 = 0$, we cannot use the exact equivalent expression for the new displacement. Instead, we must right away use the average velocity or (less accurate) the velocity at the end of the step: $x_2 = x_1 + v_2 \cdot \Delta t$.

3. Now that we know the approximate values for the position and velocity at t_2, we can proceed in just the same way as in step 2. With every new step, you propagate position and velocity forward until you reach the end of the allotted time—and, yes, you do stop there.

As already mentioned, this recipe will lead to errors in both predicted final position and final velocity. However, by using average values in the step propagation and shrinking the intervals, these inaccuracies will decrease. The power of calculus consists in the mathematically provable property that the error will be zero in the limit of shrinking the time interval Δt to zero. In this limit, intervals of any quantity—not just time—are said to be infinitesimal[9] and will be denoted by the letter 'd' rather than 'Δ'.

---

[9] Full disclosure: infinitesimal changes or differentials are not ordinary mathematical functions and we should treat them with proper respect and courtesy. Having said that, I will probably be negligent, and ask mathematically inclined readers for leniency.

Infinitesimal changes are called differentials. So the infinitesimal change dv of velocity during the infinitesimal time interval dt due to an acceleration a at some moment of time t, can be written as d$v = a \cdot$dt. Consequently, the instantaneous acceleration a is the ratio of the two differentials dv and dt:

$$a(t) = \frac{\mathrm{d}v}{\mathrm{d}t} = \underset{t_2 \to t_1}{\text{limit of}}\left(\frac{v_2 - v_1}{t_2 - t_1} \right) = \underset{\Delta t_2 \to 0}{\text{limit of}}\left(\frac{v(t + \Delta t) - v(t)}{\Delta t} \right).$$

The machinery of calculus will not be of concern to us in this book. Occasionally, I will use results without giving proof. In general, it is sufficient to appreciate that instantaneous rates of change are well defined and have specific values that are usually finite. Newton's genius is certainly not restricted to the invention of calculus, but that creation in itself was a formidable achievement. It enabled him to compare his theory with experimental and observational data—a central aspect of all of modern science.

Here, then, is the precise manner in which Newton introduces a new aspect into the concept of the temporal 'now'. Instantaneous quantities—such as velocity or acceleration—are the result of shrinking the ratio of finite differences to a 'point', the mathematical limit of contracting the denominator of the ratio to zero. In particular, any rate is the limit of a finite difference ratio with a time interval as the denominator. Therefore, the expectation that theoretical models using calculus can mimic reality is equivalent to the assumption that we can divide time into shorter and shorter intervals without limit. While this sounds reasonable and may even be true, for any physical property of matter the corresponding statement is incorrect. As an example, take the amount of a given substance—say a piece of wood. Use your finest axe and chop the piece into splinters. Then take a sharp knife and carve away. Collect the finest chips or better yet some of the dust you produced. Under a microscope, you may see a few tiny slivers that are still wood. However, if you go further, you will notice that you introduce a fundamental change. For starters, wood is a conglomerate of substances that eventually separate into different molecular structures. With the help of refined tools, you can even tear the molecules apart. Being a physicist, you are inclined to claim that from the beginning the wood was only a collection of atoms and so the division has as of yet *not* produced anything different. It will, though, when you start splitting the atom, and we all know that you can. Here is another example showing that after a finite number of divisions the character of the chopped-up quantity actually changes. Forgot to put the fabric softener into the dryer? You know you did when the sock you take out clings to the shirt or your hand because of static electricity. A standard axe will not do, but there are ways to break up electric charge, or at least to look for its natural denomination. Robert Millikan (1868–1953) was the first to do so and to find that (1) a smallest unit exists and (2) all electrical charges are integer multiples of this *elementary* charge. Today, an extremely precise value for this charge exists (if you must know: 1.602 176 6208·10^{-19} in SI units, with an uncertainty of about six parts in a billion). Present experimental evidence, as well as what is currently the best theory of matter—the standard model—agree that electrons and thus their charge are indivisible. Of course, the indivisibility of charge or the change of character when splitting the nucleus or any other example does not preclude that space and/or time can be continuously divisible. We do not know. For practical purposes, it also does not matter, because we do know that we *can* split the fleeting second into many, many fine segments whose duration we can measure with exquisite precision. Thanks to atomic clocks and other developments (see chapter 3), we now have the ability to determine time intervals as short as attoseconds—one billionth of one billionth of one second—and there is no evidence of time changing its character. Whatever that would mean, anyway. Similar fundamental questions (see chapter 6 for a few examples) are sometimes questions with vague or possibly even meaningless answers. Time is a slippery 'thing'.

1.7 A bit of mechanics—or why the solar system is stable

What is it that keeps the rhythm alive? What is the magic that keeps the grand cosmic clock running for billions of years? It appears that after the youthful universe went through a phase of massive inflation (see chapter 6), it has become thrifty and now spends only what it owns—at least to the degree that we can count the coinage. That type of currency comes in several denominations: energy, linear momentum, angular momentum, electric charge and some attributes that are more exotic. For closed systems, the overall 'amount' of these properties is 'conserved', i.e. the total sum of the values for the individual parts of the system is unchanged over time. It is not quite clear what happens in black holes, but otherwise the Universe is a closed system. Subunits of the Universe, like our solar system or the Milky Way, are only approximately closed and change does occur. However, as long as the exchange of energy, momentum, etc is not large compared to the amount present, the time scales for change measure into the millions of years and more. Of course, at certain times, for example when a supernova marks the end of the life cycle of a star, change is drastic and swift.

How can a planet slow in the rotation around its own axis or in its orbit around the Sun? One of the consequences of the laws of motion is that a force acting perpendicular to the direction of motion deflects the object without changing its speed. Conversely, any force that is not exactly perpendicular will either increase or decrease the speed. Only for elliptical orbits with the gravitational attractor, the Sun, in one of the focal points is the force component tangential to the orbital motion in just the right way to keep the orbit stable. If planets were to experience any additional force opposite to their direction of motion, commonly referred to as drag or friction, they would slow down, spiral closer to the Sun, and eventually crash— just as a low-orbit satellite exposed to a planet's atmosphere does. The vast majority of dust, debris and chunks of matter that existed in the formation period of the solar system were absorbed long ago, mostly into planets and asteroids. By now there is just not enough stuff left over to influence the planets significantly. Another potential source for drag could be the corona and the plasma halo of the Sun. Yet even Mercury as the planet closest to the Sun is in no danger of crashing into the Sun any time soon.

Even so, the stability of the planetary orbits is not mathematically absolute, mostly because of the gravitational pull of large planets like Jupiter and Saturn and/ or of their own moons, which are much smaller but also much closer. These gravitational forces are acting along lines that at any given moment do have a component that is tangential to the planet's orbit. In the long run, though, this direction is averaged out and does not lead to a long-term net increase or decrease of the speed, only to quasi-periodic variations of the orbital and rotational parameters. For Earth, the so-called Milankovitch cycles are an expression of the complexity of these variations. Leading terms of the overall variation include the already-mentioned precession of the rotation axis, as well as a change of the eccentricity of Earth's orbit. Typical periods for the dominant variation are tens of thousands of years, but all changes combine to convoluted and aperiodic motion. Overall, then, the length of the sidereal periods of planetary orbits are nearly unchanged, since they only depend on the semi-major axis, which at least to first order is unaffected by the perturbations. Thus, while running in a slightly bumpy way, the great clock in the sky runs true and for a long while without the need for rewinding.

1.8 Not quite a clock yet—the Antikythera mechanism

In 1901, a small Greek ship carrying sponge divers back home from a not-so-successful stint in the Western parts of the Mediterranean Sea was caught in a storm. The crew sought shelter at the coast of the small island of Antikythera north of Crete. When the divers decided to explore the underwater neighborhood, they did not find sponges. Instead, they discovered what turned out to be the remains of a ship that had fallen victim to another storm about 2000 years earlier. The find created quite a stir in archeological circles, since it was the first discovery of an ancient shipwreck. The cargo scattered on the bottom of the sea included sculptures and glass vessels of various kinds, giving valuable clues for the date of the ship's journey, which was estimated to be around 150–100 BCE. Among the scattered parts on the seafloor, the divers found fragments of a geared mechanism that archeologists later proclaimed to be an astrolabe—an instrument used to measure vertical angles for navigation and stargazing. Soon afterwards, the crated pieces were stored in the Museum of Athens, and forgotten almost for good. Half a century later, a careful reexamination [10] led to the startling insight that the mechanism was actually a sophisticated mechanical device capable of keeping track of a variety of periodic events—both on Earth and in the heavens. The machine allowed calendric calculations, such as determining the times of the Olympiad. Even more astonishing, this analog 'computer' was capable of mimicking the motion of the Sun and Moon, including the prediction of eclipses. Possibly, although this is unclear, the device could also show the position of the five planets known to the Greeks. Nothing even remotely resembling such a mechanism was (or is) known from that early time. The level of sophistication of the overall design and construction, in particular of the gears, and the inherent precision of this earliest known astronomical simulation mechanism were unmatched until the fourteenth century. Nevertheless, this remarkable instrument was not a clock. It did not produce repeatable, constant units of time, and neither did it otherwise allow measurement of the duration of elapsed time. There is also no known later mechanism that could be seen as has having benefitted from this early invention. On the other hand, there is something transcending the uniqueness of this machine. Like Stonehenge or Ptolemy's crystalline spheres, the Antikythera mechanism is an attempt to reproduce and capture at least aspects of the natural rhythms surrounding us. Using purely mechanical means, it acknowledges that such rhythms exist and succeeds in mimicking them. To my mind, this is the general hallmark of dynamic model building and specifically an engagement with the notion of temporal change. The following chapter investigates the next steps of this journey, with a focus on the mechanical devices that we properly call clocks.

References

[1] Hartmann W K and Davis D R 1975 Satellite-sized planetesimals and lunar origin *Icarus* **24** 504–15

[2] Šprajc I 2017 Astronomy, architecture, and landscape in prehispanic Mesoamerica *J. Archaeol. Res.* https://doi.org/10.1007/s10814-017-9109-z

[3] Barnhart E L The First Twenty-Three Pages of the Dresden Codex: The Divination Pages

[4] www.space.com/21719-vega.html
Sharrah P C 1975 Pole stars of other planets *J. Arkansas Acad. Sci.* **29** 62–3

[5] http://image.gsfc.nasa.gov/poetry/ask/a11846.html

[6] Kuhn T S 1957 *The Copernican revolution: Planetary astronomy in the development of Western thought* (Cambridge: Harvard University Press)

[7] Koenib L R *et al* 1967 Handbook of the Physical Properties of the Planet Venus NASA SP-3029

[8] The homepage of the JPL is a good starting point to explore: http://ssd.jpl.nasa.gov/? ephemerides

[9] www.quartets.de/acad/firstlaw.html
Pask C 2013 *Magnificent Principia: Exploring Isaac Newton's Masterpiece* (Amherst NY: Prometheus Books)

[10] de Solla Price D 1974 Gears from the Greeks. The Antikythera mechanism: a calendar computer from ca. 80 BC *Trans. Am. Phil. Soc.* **64** 1–70

Chapter 2

Hours, minutes and seconds

We live in busy times with a concurrent need to tell the time in ever more precise ways—or at least so we are told. The urgency of that need (perceived or real) and its wide reach may be modern, but already two millennia ago some citizens of Rome complained about the ubiquity and rushing influence of clocks in the public sphere. To be sure, those clocks were sundials and the acceleration of the pace of life they created was presumably tame by our standards. Our clocks are certainly more precise. Still, the idea is the same. The concept of dividing the day's length into smaller pieces is indeed old and a handful of related tools are more than 3000 years old. The changing length of a shadow cast by a stick is all there is to the principle of a sundial. The changing amount of sand or water due to steady in- or outflow is all there is to the concept of the hourglass or the clepsydra. All three timing tools— sundial, hourglass, water clock—rely on continuous motion that does not offer any intrinsic, 'natural' subunits. How it happened that 12 hours were used to add up to the length from sunrise to sunset (and 12 more hours to include the night) is not completely certain. An important clue might be had from the number 12 itself, as it is divisible by 2, 3 and 4. A bit of additional numerology, namely the extra factor 5, can be invoked for the number 60, used to subdivide the hour first into minutes and then seconds[1]. Possibly also relevant is the relation of the number 360—as the number dividing the circle into equal parts or degrees, a convention usually traced back to Babylonian astronomers—to the division, even if only approximately correct, of the year into 360 days. If the year was thought of as being due to the circular course of the Sun, then this correspondence between the geometry of the

---

[1] Reading up for this book, I learned how to count, with the fingers of your two hands, up to 60 in a 12-based system—surely a long known fact, just not by me. I had always thought that our ten fingers (including the thumbs) predestine us for the decimal system. Not so fast. Exclusive of the thumb, each of your fingers is made of three segments. By pointing the thumb sequentially to these segments on one hand, you can thus count to up to 12. Fold first the thumb and then the four fingers of the other hand for each completed dozen and you can keep track of numbers up to 60. It is really easy.

circle and measurement of time might not be a coincidence. In any case, during the reign of pharaoh Thutmosis III (1501–1447 BCE), sundials that divided the time of sunlight into 12 units were already in use [1] (figure 2.1). No matter how and when it started though, the choice of having 24 hours in a day, 60 minutes in an hour and 60 seconds in a minute *is* a convention—an old and stubborn one.

2.1 Origins of the metric system and the SI unit of time

Fast forward to the French Revolution, one of those events in human history whose repercussions are felt far and wide. The excesses of the revolution and its aftermath go along with important contributions to social and political progress. Part of that advancement was the introduction of rational standards for conducting business, including a unified system to quantify goods and services—the metric system of units. One goal was to base the new units on building blocks found in nature rather than on the whim of a local authority or the heel-to-toe length of the resident duke's foot. For the length unit it was decided to make it equal to one ten-millionth of one quarter of Earth's circumference, i.e. of the distance between pole and equator. Our meter is truly a measure of the Earth. While the revolution was still in full swing, expeditions were sent out to the north and south of France to perform the necessary precision land surveys and triangulations [2]. In the end, the meter thus obtained turned out to be short by 0.2 mm or two parts in 10 000. I think this level of accuracy is quite an accomplishment. There are several reasons why the measurement was not the intended exact integer fraction of the quadrant of the meridian. What does it actually mean to express the length standard in terms of a certain distance along the surface of the Earth? By 'surface' the new definition of the meter obviously did *not* mean Earth's topographic surface, following each hill and valley. The line, the meridian from pole to equator, was thought to follow a smooth, idealized surface at a constant height, say at sea level. Therefore, the triangulation needed to and did include the full three-dimensional structure of the lattice of all the measured triangles. We have known since Newton that Earth's rotation around its polar

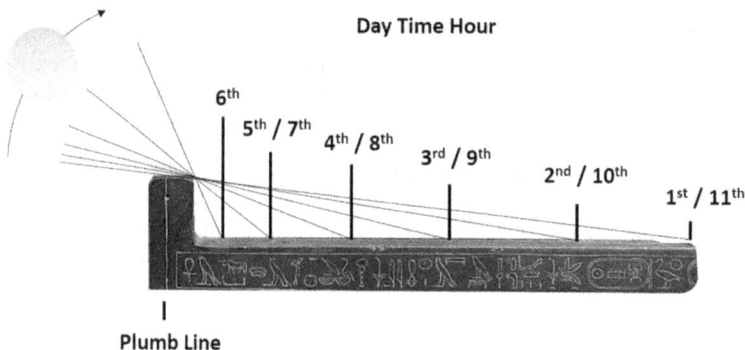

Figure 2.1. Sundial from the time of pharaoh Thutmosis III (1501–1447 BC (Museum of Egyptology, Berlin, Germany) showing marks to divide daylight period into 12 units. The schematic illustration indicates use of the device and origin of uneven distance between the counting notches.

axis induces a bulge at the equator. Therefore, when the surveying data were analyzed the actual three-dimensional surface had to be mapped onto an appropriate ellipsoid. Unfortunately, this step of the calculation contained a flaw, which explains why the meter is slightly off-target. Another cause lies in the mathematical complexity of how the myriad of measurement uncertainties are reflected correctly in the final result. Today, we would add another challenge. The very idea of a smooth geometric shape—whether spherical or not—is flawed. There is no such thing as a well-defined single pole–equator distance (or equator circumference or any other measure). Earth's ideal surface, the geoid, is everywhere perpendicular to the direction of gravity. Because of the irregular, lumpy distribution of mass in the interior of the Earth, the actual shape of surfaces of constant gravitational acceleration are highly irregular, and because of Earth's interior dynamic even change with time. A few lessons can be drawn from this story. The advancement of basic and applied science go hand in hand; claims of perfection and finality should be viewed with caution; it is important to go forward in the face of imperfection. We owe to the French Revolution a powerful and universal system of measures based on clear definitions and the decimal numbering system. Except for the unit of time.

Since ancient times, people have used the seven-day week, the 24-hour day, the 60-minute hour and the 60-second minute. For a radical advocate of the French Revolution that in itself might have been reason enough to look for alternatives. Be that as it may, taking advantage of the decimal system a simple, consistent definition of new time units was indeed easily devised (see figure 2.2). Divide the day into 10 metric hours of 100 metric minutes each. Each new minute in turn comprises 100 metric seconds. That gives 1000 metric minutes in a day compared to our customary $24 \cdot 60 = 1440$ minutes. Thus, the newly invented metric minutes were about 44% longer than their pre-revolutionary counterparts. Likewise, since 100 000 'new' seconds fill one day, compared to $24 \cdot 60 \cdot 60 = 86\,400$ 'old' seconds, the former was about 14% shorter than the latter. These differences are certainly substantial and noticeable, but they do not really explain why the decimal hour did not catch on. After

Figure 2.2. Decimal clock from the French revolution.

a brief experiment, France abandoned the metric hour and reverted to the age-old ways of dividing the day into smaller parts[2].

Within the international metric system, the second is the only unit of time and the standard practice of multiplication and division by 1000 is the only way to modify the base unit. But it is highly impractical to denote, for example, the length of one day by 86.4 ks or the length of one year by 31.557 Ms. Thus the properly defined non-metric units of year (a, for the Latin *annus*), day (d), hour (h) and minute (min) are being used alongside the second.

About 50 years ago, in 1967, our present-day definition of the second was introduced as a certain number of periods of a well-characterized oscillation of the cesium atom. To be precise, one second is now defined as 9 192 631 770 oscillations of the electro-magnetic radiation emitted by cesium atoms when they switch between two particular electronic states (more on that in chapter 3). While the incarnation is thoroughly modern, the idea itself of marking off short periods of time is old—maybe inspired by our heartbeat or the blink of an eye. The *helek* (Hebrew, חלק), for example—used in the Hebrew calendar—is such a short time unit. It is derived from an even earlier Babylonian time-reckoning measure, equivalent to 1/72 of 1° or 1/72 of 1/360 of the duration of one full celestial rotation. If we count that duration as 24 hours and each hour as 3600 seconds, then the *helek* comes out to be 3⅓ seconds. Since about at least the year 1000—when al-Biruni mentions the second in his *Vestiges of the Past*—until 1960, our fleeting second was identified as a certain fraction (1/86 400 = 1/(24·60·60)) of the solar day. As we saw in the previous chapter, the *actual* time from noon to noon varies and so an *average* length of time was introduced, and it was left to astronomers to clarify what the latter is. For seven years, from 1960 to 1967, the reference for the second shifted to the more exactly characterized orbit of the Earth around the Sun. All in all, up until the last decades of the 20th century, measurement of time was connected to the natural rhythms our senses can easily detect. Nevertheless, even in this context increasingly precise definitions were ever more removed from direct experience, or, as Dohrn van Rossum pointed out, 'what appears in retrospect as a long term development toward greater temporal precision amounted to within the purview of contemporary experience above all to many steps of abstraction' [4]. In that spirit, we will first ask what the general key parts of a clock are before we return to a discussion of specific aspects of measuring time.

2.2 A generic clock

What makes a clock tick? The answer depends, of course, on the specific clock that is ticking—if it ticks at all. Later, we will explore the inner workings of clocks with more specificity. For the moment, the question is simply which ingredients do we need in general to construct a clock. Figure 2.3 summarizes them for the case of a

---

[2] It is fascinating that decimal-based time reckoning in China started much earlier and was much longer lived than in Europe [3]. For well over 2000 years, the day was divided into 100 units, called ke. Only in the late 17th century, following the adoption of European astronomical theories and practices, was the length of the ke decreased by 4% so that 96 ke fill one day and thus four ke make up one hour.

time keeper

{clock} = {power source} + {movement} + {gearing} + {indicator} + {counter}

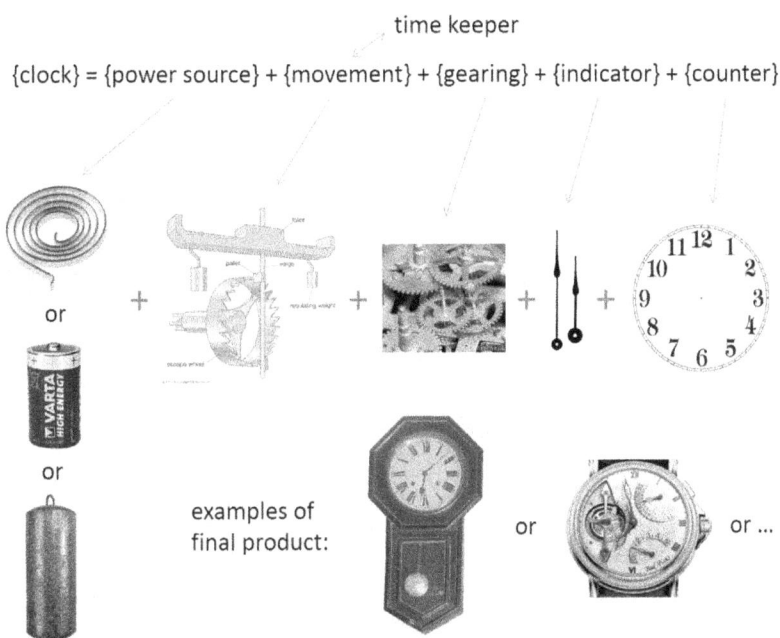

or

examples of
final product:

or

or ...

Figure 2.3. Generic ingredients of a generic clock. Apart from the energy source needed to overcome friction, an oscillating timekeeper and mechanisms to count and indicate the number of periods elapsed are needed. Often, gears are also necessary to enable proper counting if the movement is too rapid for the units of time to be indicated by the clock.

mechanical clock, but you can find the functional equivalent of each part in most any type of time-telling instrument. The central item is the movement, the time-keeper that in most cases utilizes a repetitive process, though there are exceptions[3]. Because of unavoidable *friction*, in the general sense of energy loss, a power source will be required, feeding energy back into the oscillator at just the right rate. Time-keeping and time-telling are two different things, and the hardware needed to count, indicate and possibly record the number of elapsed temporal units is an essential part of any clock. The link between time-keeping and time-telling is shown in the figure by a gearing mechanism, which also allows the slowing down or speeding up of the time indicator relative to the timekeeper. While in principle dispensable, gears are a ubiquitous part not only in mechanical, but also in atomic clocks, where frequency multiplication is essential, as we will see in the next chapter. A generic clock, then, consists of at least these three items: an energy source, a timekeeper (typically but not always an oscillator) and a counter/recorder. The type, design, choice of material and many other factors—not least style and artistry—all play crucial roles in the

---

[3] Burning of incense strings, flow of water or sand through narrow openings, and radioactive decay of unstable nuclei are examples of non-repetitive processes that have been used to measure time intervals. By the way, even clocks built around a linear, non-cyclical timekeeper incorporate the above list of key parts. The hourglass, for example, needs gravity as the power source, uses a timekeeper (the steady flow of sand grains) and keeps track of the elapsed time by the volume of the sand accumulating in the lower bulb.

manufacture, durability, cost and quality of any time-telling apparatus. The variety of real clocks is astonishing. Not many other pieces of machinery have found as much appeal and such a wide range of application for such a long time as clocks and watches.

These days, a typical quartz watch gains or loses a second every day. Invest more money (ready to buy an atomic clock?) and that accuracy improves dramatically. However, nothing in this world is 'perfect'—which some take to mean that everything is just the way it is and thus everything *is* perfect. Here is not the place to argue about these views, other than to say that I will use the word 'perfect' only to point out a comparison with certain mathematical models, usually characterized by simplifications such as the complete absence of friction. On the other hand, physicists are interested in the real world, 'blemishes' and all. Physics is also all about connecting ideas with measurement and vice versa. In our context, this means we are interested, for example, in questions like why different types of clocks run more or less 'smoothly' than others. As a first step to quantify such variability, the following procedure is useful: let a group of clocks, built to the same specifications, run for a while under identical conditions and then determine the spread of the time the clocks indicate has elapsed. We find that pendulum clocks do much better than hourglasses and atomic clocks outperform pendulum clocks by a vast margin. Of course, there is more to the quality of clocks than this simple criterion: lousy clocks can be off by the same amount and still be lousy clocks. We will return to the interesting and tricky question of clock comparison in chapter 3. Beyond characterizing the quality of clocks, we are also interested in the cause of these variations in order to possibly build a better clock. Colluding with mathematicians, physicists have also found a way to quantify the idea of 'perfection' so that real clocks can be stacked up against ideal ones, even if the latter only exist in theory. The following sections of this chapter explore these ideas using simple mechanical 'toy' clocks that differ in their time-keeping mechanisms. Specifically, we look at three gadgets that dispense with any unnecessary detail: (1) a bead bouncing back and forth between two rigid but elastic walls; (2) a vibrating object attached to an elastic, coiled spring; and (3) a pendulum. Together they encapsulate the conceptual base of repetitive time-keeping devices. For quite a while, the best *real* time measurement device available was based on the pendulum clock and the end of this chapter is devoted to a discussion of this clock. Only in the second decade of the 20th century did first quartz and then later atomic oscillators replace purely mechanical clocks[4].

In the complete absence of friction in the widest sense, either one of these mechanisms, once set into action, would repeat the very first complete path of its motion exactly. Attach a counter mechanism—e.g. a light beam either reflected or obscured by the oscillating mass or the vibrating pendulum bob—and you have a clock whose precision is limited only by the accuracy with which the return to a

---

[4] Up until the late 1920s, the National Bureau of Standards (NBS, the predecessor of the current keeper of time in the USA, the National Institute of Standards and Technology, or NIST for short) used sophisticated pendulum clocks as the US time standard. The last of such pendulum clocks had a precision of about one millisecond per day when it was replaced by a quartz oscillator-based standard [5].

given position can be measured. This statement is correct to the extent that the environment of the oscillator can be kept constant and materials do not age or change. In such a case of an oscillator in well-controlled surroundings, the period of the timekeeper is constant to the extent that friction and noise can be kept to zero. As an integral part of any measurement, the influence of external effects—be they systematic or random—needs to be accounted for. How we can do that is also part of this chapter.

2.3 Toy clock I: bouncing bead in a box

Probably the simplest mechanical clock conceivable (but not one you can actually manufacture) is the following device. Take a rigidly mounted box[5] of length L, and place a smooth round ball of radius R somewhere into the box. Now give the sphere a little kick so that it glides with speed v parallel to one of the edges of the box and towards the right side. The box is made of a special material that is perfectly stiff and does not absorb any energy from the ball upon impact. Since no energy is lost, the ball bounces back and reverts its direction of motion while the speed stays the same. After a certain time, the same happens on the left side of the box with the ball now being kicked to the right and so forth. A full period is given by the time needed for the bead to return to the same location with the same direction of motion, e.g. going past the midpoint of the box from left to right or just touching the left wall and being instantaneously at rest. This period, customarily denoted by T, is determined by the time it takes the bead to travel the distance $2L_{eff}$ where $L_{eff} = L - 2R$ as can be seen in the insert of figure 2.4. Since the speed v is constant, we find that $T = 2L_{eff}/v$. In the limit of no friction, an uninterrupted series of soundless ticks (or maybe clonks) will ensue, with each ball–wall encounter separated by the constant time interval T, the period of this particle-in-a-box clock.

It is instructive—and useful for discussions in subsequent sections—to represent the repetitive motion in graphical form. Figure 2.4 displays the position of the bead along its direction of motion (the dependent variable, plotted along the vertical axis) as a function of time (the independent variable, plotted along the horizontal axis). Rather than using specific numerical values, it is convenient to scale both position and time in relation to relevant quantities. Here, the position is given in units of one-half of the effective box length L_{eff}. Thus, the position of the right wall is represented by $(x = 1/2L_{eff})/(1/2L_{eff}) = 1$, that of the center by 0 and that of the left wall by -1. Likewise, time is scaled by the period of one complete round-trip. Thus, in this unit, the time it takes to travel between the two walls is ½, and the travel time between the box center and a wall is ¼. The black solid line represents a situation in which the particle was initially at rest at the center of the box ($x = 0$) and then, at time $t = 0$, received a kick that sent it towards the right wall. Afterwards, the bead bounced

---

[5] Okay, so the box should really be more like a tube. But then my alliteration would have been out the window. Besides, we will encounter a similar contraption both in quantum mechanics (with the name of a 'particle in the box', see chapter 3) and in special relativity (with the bead replaced by a light pulse, see chapter 4). By the way, a completely equivalent arrangement looks like a single-bead abacus: a sphere of radius R that glides frictionlessly along a straight wire of length L mounted between two immovable walls.

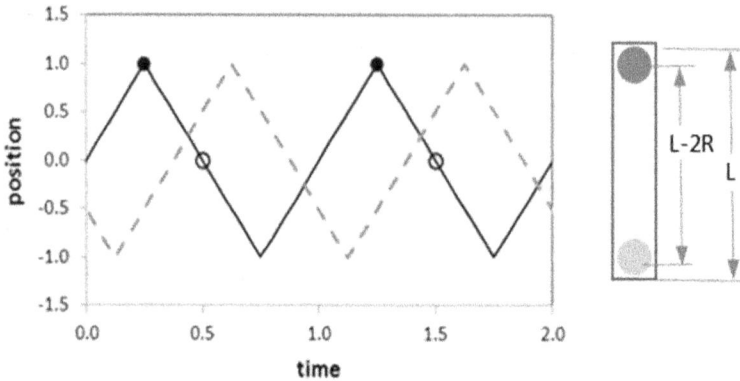

Figure 2.4. Graphic representation of bead position (vertical axis) with respect to time (horizontal axis). The motion of the bead for two different initial conditions, described in the text, are shown. The period of the repetitive motion is given by the horizontal distance between a pair of equivalent points (two examples are indicated).

back and forth between the two walls at $x = +1$ and $x = -1$. In-between, the speed of the particle was constant, so that the position changed linearly with time. The period T of oscillation is the time between two equivalent points along the trajectory—two examples are indicated in the figure by the pairs of filled and open circles, respectively. Also included is another case (dashed trace) in which the particle was initially placed halfway between the box center and the left wall and was then, at $t = 0$, set into motion with the same speed as before, but now towards the left. As can be seen in the figure, the period is unchanged, as it should for a particle moving with the same speed. The shift between the two cases can be viewed as an offset in the starting point of two otherwise identical clocks. Because the two cases assumed the same strength for the kick that got the bead and thus the clock moving, the slope of corresponding linear portions are the same. To put it differently: the speed of the particle depends on the strength of the initial kick. In the graph this will show up as a changed slope of the linear portions—steeper for a stronger kick and faster motion, and less steep for a softer kick and slower motion—which of course will alter the period of the clock.

2.4 Friction I: spoiling the bouncing bead

Because no energy is lost, the traces in figure 2.4 repeat unchanged for arbitrary times. In real life, however, some friction in one form or another will remain despite our best precautions, and the ball will slow down somewhat on its way to the next wall encounter and/or due to its interaction with the walls. An example of such a situation is depicted by the solid line in figure 2.5 for the case of energy loss occurring only at the walls (the dashed line shows the friction free case for comparison). Not all is lost, though, because we can imagine a clever mechanism that senses the actual speed of the ball each time it makes contact and then provides a kick so that the ball accelerates to make up for the lost time on its way in. No matter how clever that mechanism, it cannot be perfect, and the resulting speed after one feedback will differ, up or down, by an average amount δv from the design value v. This leads to

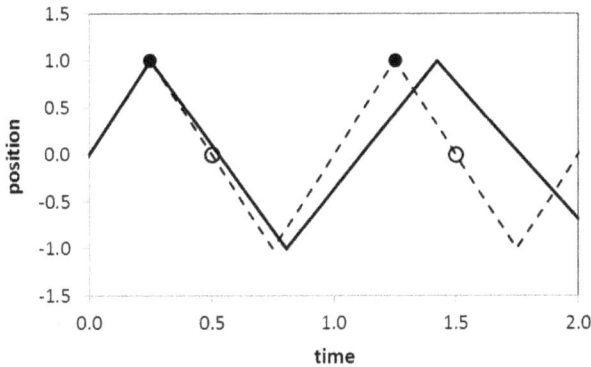

Figure 2.5. The graph illustrates how friction leads to a slowing down of the bead and thus to an increasing period defined, for example, as the time between two subsequent encounters with the same wall (filled circles) or between two subsequent equivalent mid-point crossings (open circles).

an average change δT of the next period in the same proportion. These changes can affect our clock in two ways. If the changes δv and thus δT are truly random, then their averages are both zero, $\langle \delta v \rangle = 0$ and $\langle \delta T \rangle = 0$, and at least in the long run the clock will run perfectly fine. However, the length of each period fluctuates by the same average percentage as the speed does, i.e. by the amount $|\delta T| = T \cdot \delta v / v$. Furthermore, if our feedback mechanism is imperfectly calibrated, leading to a biased speed change—either a bit too much or a bit too little—then the long-term averages are non-zero. After a time $t = NT$ has elapsed, the clock will be off relative to a perfect clock by an amount $\Delta t = N \cdot \delta t = (t/T) \cdot \delta T \neq 0$. For example, if a box clock with period of one second ($T = 1$ s) is to be off by not more than one second ($\Delta t = 1$ s) per day ($t = 86\,400$ s), the bias must not be more than $\delta T = \Delta t \cdot (T/t) \approx 1.2 \cdot 10^{-5}$ s. In other words, both the period and particle speed are expected to be known and held constant to within about one part in a hundred thousand. Quite a challenge!

If it is difficult to engineer such precise feedback, maybe we can dispense with that need and concentrate on eliminating damping as much as possible. Early clock-makers were also material scientists, exploring different materials, but they were also not shy to use grease—sometimes lots of it. Suppose that no feedback is provided for our bouncing bead and that per trip between two walls an average non-zero speed loss Δv occurs. Furthermore, assume that this change is always the same fraction α of the average speed on any given segment. In equation form: $\Delta v / v = \alpha$. While this relation is, strictly speaking, only a tractable mathematical model, it does provide a reasonable approximation to real behavior[6]. Figure 2.5 captures the essence of this situation. Note that because the bead slows down, the slope in the graph decreases in proportion. Correspondingly, the period of the clock increases relative to the ideal case shown by the dashed curve. In reality, the slope of each segment will change continuously while the graph substitutes this behavior by an abrupt, equivalent

---

[6] For example, a free-falling object subject to a drag force of this nature will reach a terminal velocity in agreement with observation.

change of the speed at each bounce at the wall. What will be the speed v_N after N wall collisions? With an eye towards a more quantitative description, we observe the following. After a single collision, the speed is diminished from the original value v by the amount $\Delta v = \alpha v$, i.e. $v_1 = v - \Delta v = v - \alpha v = v(1-\alpha)$. After two collisions, we similarly find that $v_2 = v_1 - \alpha v_1 = v_1(1-\alpha) = v(1-\alpha)^2$. In general, then, after N wall collisions we have $v_N = v(1-\alpha)^N$, with a corresponding period given by $T_N = L_{eff}/v_N = T/(1-\alpha)^N$ where we have used the above relation $T = 2L_{eff}/v$. Provided that the speed change is small ($\alpha \ll 1$), the last equation can be approximated[7] by $T_N \approx T \cdot (1+N \cdot \alpha)$. Therefore, after N wall encounters the period has changed from T to T_N, or roughly by $\Delta T = T - T_N \approx \alpha N T$. This relation can be used to estimate the largest allowable fractional speed change α in order to maintain the period constant to within ΔT, namely $\alpha \approx \Delta T/NT$. Because the period of our clock decreases, the number N of ticks marking off one day is now larger than before and 86 400 is only a lower limit. Thus, in this example of a desired time measurement stability of one second per day, an upper limit is $\alpha_{min} \approx 1/86\,400 \approx 1.2 \cdot 10^{-5}$, the same value as found above for the necessary feedback precision. It is certainly not easy to construct a mechanical device—ball in a box, mass on a spring, pendulum, anything—with that level of consistency of motion without feedback.

The box clock is a fine timekeeper in principle, if not in practice. However, it does have a fundamental drawback. Nothing in its physical make-up determines the duration of its period: for a clock of length L, the period is determined by the speed v and thus by the initial kick given to the ball. Since the speed can have any value, so can the period. In other words, the period is not an intrinsic property of the clock. Therefore, it would be very difficult to compare two such clocks: the bouncing bead is not really a good clock candidate after all. There is one exception: substitute the material bead by a light pulse. In that case, the speed of propagation is unchangeable. We will explore such a light clock in chapter 4.

Before we turn our attention to the next prototype clock, there is one more lesson to learn from our current toy model. Because we have acknowledged that friction or damping is unavoidable, we need to find a better way to quantify its influence on clock performance. For this purpose we introduce a parameter that is relevant for all types of oscillators, the so-called quality factor Q. This quantity is a measure for the degree of slowing down ('decay') due to damping and in the absence of any feedback. Explicitly, Q can be defined (there are other definitions) as the ratio between the time τ during which the oscillator loses a certain amount of energy and the time T needed for a single period of oscillation, i.e. $Q = \tau/T$ or $\tau = QT$. Note that Q is a pure number without units and indicates how many oscillations occur before the prescribed amount of energy is lost from the system. The larger the value of Q, the less damping there is per oscillation period or, conversely, the more oscillations occur before the energy is lost. A system with high Q is a system of high quality. For well-designed and manufactured mechanical devices, quality factors of the order of

---

[7] On occasion, I will use mathematical language, ignoring Stephen Hawking's (I think it was him) warning that every equation reduces the number of readers by 10%. Even with only 10 equations, the outlook is dim. But I am an eternal optimist.

1000 or more are possible. However, as we will see later, atomic clocks feature Q factors of the order of 10^{12} and more, which is impossible to achieve with mechanical clocks.

Obviously, the specific length of the decay time τ—and thus the value of Q— depends on the amount of energy that is disappearing per cycle. For the present purpose, we identify τ somewhat arbitrarily with the length of time needed so that both the speed of the particle in the box and the period of the clock decrease by a factor of two. In other words, after a time τ, the original period T has been reduced by $\Delta T = 1/2T$. Incidentally, the kinetic energy, being proportional to the square of the speed, will have dropped by a factor of four. As was pointed out above, the change of period ΔT after N vibrations with a fractional velocity loss of α per round-trip is given by $\Delta T \approx \alpha NT$. Applied to the current context ($\Delta T = 1/2T$), we find that $N \approx 1/(2\alpha)$. Since the decay time can be expressed in terms of this same number of oscillations, we finally have $NT = \tau \approx T/(2\alpha)$, so that the Q factor may be estimated by $Q = \tau/T \approx 1/(2\alpha)$. Using the above value for α, the Q factor of about $1/(2\alpha) \approx 10^5$ follows. Such a system would lose one second per day without any feedback. Another useful way to think about the significance of the Q factor emerges from the realization that if $\tau = NT$ then the definition of the quality factor immediately implies that $Q = N$. Recall that N is the number of oscillations needed to lose a given amount of energy. In principle, this relation provides a method to determine the quality factor experimentally: count the cycles it takes for the velocity to slow down to a given fraction (say one-half) of the initial value. The line of reasoning presented here is an example of order of magnitude assessments or ballpark figures that can be quite useful at times. We will sharpen our understanding of the parameter Q as we go along, but for now the take-home message is that quality factors of the order of 100 000 may be needed to have free-running clocks of the precision of the order of one second per day.

2.5 Toy clock II: the simple harmonic oscillator

Our next prototype clock is built around a device known in physics as a harmonic oscillator. In its idealized version, such an oscillator consists of a small object with mass m attached to one end of an elastic spring of length L and negligible mass. The other end of that spring is clamped to an immobile support point. Kick the object or pull or push it a distance away from its resting point and it will start to vibrate. As much fun as it would be to kick the mass sideways and watch it go through some pretty nifty gyrations, we will exercise some restraint and only pull, push and kick the object along the spring axis. That way, we have one-dimensional motion and the mathematical description will be much simpler.

In our first toy clock, the free movement of the particle-in-the-box lacked any fixed, intrinsic time scale. Precise time-keeping has to be actively ensured by a complex feedback mechanism. Now that the motion is no longer force-free but guided by the spring, is there a time scale that is characteristic of the system? Assume that the mass is initially resting at the midpoint. When it is kicked, it will move a certain distance in the direction of the kick before the spring force pulls it back.

Because the length of this excursion, i.e. the amplitude of the oscillation, necessarily depends on the strength of the initial kick, it might seem that here as well the period of oscillation depends on how the system is set into motion. However, the harmonic oscillator has the amazing property that it will spend exactly the same amount of time for one full oscillation regardless of how hard it was kicked. In other words, the period of oscillation of the harmonic oscillator is independent of its amplitude. A very useful feature when you want to build a clock! In the following paragraphs we will see how this property comes about.

An elastic or linear spring is a coil that behaves according to what is known as Hooke's law: stretch or compress the spring by ΔL relative to its length L and the spring pulls or pushes[8] back with a force proportional to the extension ΔL from equilibrium; the proportionality constant is usually denoted by k. This parameter is directly related to the stiffness of the spring: the softer the spring, the weaker the restoring force and the smaller the magnitude of k. Stiffness is a physical character-istic determined by the composition and thickness of the spring material as well as by the specific shape of the spring.

When the spring is stretched (say by pulling to the right), then the attached object experiences a force due to the spring in the opposite direction (i.e. to the left). The same is true for compression of the spring, so that when the spring is pushed, it pushes back on the mass. Mathematically, the opposite direction between spring force and spring length change is expressed by a minus sign in front of the proportionality constant, i.e. $F = -k\Delta L$. This minus sign is crucial for oscillatory behavior. Without the minus sign, the force points in the same direction as the length change of the spring. In this case, the spring force *increases* the spring extension, which further increases the spring force pushing the attached object further, and so on. In other words, a run-away scenario unfolds in which the mass m accelerates continuously and moves further and further away from equilibrium. Such motion requires an external energy source—which can be arranged—but it really does not lend itself as a timekeeper for the kind of clock we want. So, minus it is. On a different note, it is important to realize that gravity does not play a role in this type of oscillator, so it could run on a spaceship in permanent free fall[9], such as the International Space Station, or onboard space probes like the Voyager mission, venturing far from any source of gravitational pull.

In this chapter, we will only encounter the harmonic oscillator as a mechanical system. However, the associated concepts are much more widely applicable. For example, when we discuss atomic clocks in chapter 3, we will see that to a reasonable approximation atoms exposed to light respond as if the electrons inside the atom are attached via a linear spring to a massive object, the nucleus of the atom. Many

---

[8] Some types of shock absorbers contain springs that can be both stretched and compressed. Most other springs are made of coils whose adjacent turns touch when released. Obviously these springs cannot be further compressed. In such cases, we can 'pre-stretch' the spring to a new length, which then acts as the above equilibrium length. Pre-stretching can be achieved by attaching the mass between *two* springs whose opposite ends are attached to fixed points suitably far away from each other. The attached mass will then behave like a harmonic oscillator in the sense above.

[9] See chapter 4 for how that relates to the absence of gravity.

qualitative aspects of absorption, emission and scattering of light by atoms and molecules can be understood based on the simple, classical image of a harmonic oscillator responding to a periodic stimulus. To be sure, a detailed and quantitative understanding of salient features of the interaction between light and matter requires quantum mechanics. In fact, we will see that quantum theory was developed *because* classical physics *cannot* explain certain basic observations. We should also be clear that the notion of a linear, elastic spring is an approximation. Real objects, mechanical or otherwise, deviate from the linear relationship between restoring force and displacement from equilibrium. However, the deviation is often quite small, so that to a very good approximation the following discussion holds. Harmonic oscillation is a concept that is extremely fruitful and its wide range of applications in physics and engineering may surprise new learners.

Returning to the mechanical oscillator, our next task is to determine its period of vibration and how the period is related to the physical properties of the system. For our toy system only two such parameters exist, namely the mass of the object and the stiffness of the spring. We can assume that Newton will help. To that end, we start by inserting the specific form of force into Newton's second equations of motion (see chapter 1). For one-dimensional oscillation of a single object, the case of interest here, we only need to use the second equation, i.e. force = acceleration × mass ($F = m{\cdot}a$). By the way, one-dimensionality does *not* imply straight-line motion. The line along which the object oscillates can be bent; for example, the path might be a segment of a circle. If the line *is* straight, we will characterize the location of the moving object with the variable x, and if the path is curved—as is the case for the bob of a pendulum—we will use the label s to indicate position. Putting this together, we find that the acceleration, the rate at which the velocity of the object changes, is proportional to the distance between the current location of the object and the equilibrium point. As always and in accord with Newton's equation of motion, the acceleration caused by a given force— here the spring—is inversely proportional to the mass of the object. You can express all that with a few symbols, $a = -(k/m)x$. In chapter 1, we caught a glimpse of how Newton's equations of motion can be used to determine the path of objects influenced by well-characterized forces. For the current situation, the equation of motion takes on the following form

$$m \cdot a = F \quad \text{with} \quad a = \frac{\mathrm{d}^2 x}{\mathrm{d}t^2} \quad \text{and} \quad F = -k \cdot x \quad \Rightarrow \quad a = \frac{\mathrm{d}^2 x}{\mathrm{d}t^2} = -\frac{k}{m} \cdot x. \quad (2.1)$$

Equation (2.1) looks like an ordinary algebraic equation. However, because both the acceleration a and the position x vary with time and are dependent on each other, with acceleration being the second rate of change of position, (2.1) is a *differential equation*—as all equations of motion are. Depending on your familiarity with such mathematical relations, the above equation may look either rather simple or rather formidable. The important part is that Newton's theory of mechanical motion, together with the general mathematical theory of differential equations, allows us to analyze and predict the motion of an ideal, i.e. undamped, oscillator. Even if you are not familiar with the symbolism of differential calculus, you can hopefully appreciate that this abstract representation allows us to express content and meaning in a very compact manner. Mathematical equations are a bit like a foreign language. Or a *quipu*, a collection of knotted strings used by the Inca to count goods. *Quipu*, language, or mathematical expressions—they all combine function and form, utility and, yes, beauty and elegance.

As already mentioned, acceleration *a* and position *x* change as the mass on the spring oscillates and they are thus, mathematically speaking, *functions* of time. In a more explicit symbolism, one would therefore write the two quantities in the last form in relation (2.1) as $a(t)$ and $x(t)$, respectively. Solving the differential equation (2.1) means finding the function $x(t)$ whose second order rate of change at any moment in time is, apart from the minus sign, proportional to the value of the function itself. It sounds complicated and, depending on your mathematics background, it may or may not be. Mathematics should not get unduly in the way of our story. So, here we simply state that the function $x(t)$, which—in the above stated sense—solves equation (2.1), can be expressed as a trigonometric function, specifically as the sine function.

$$x(t) = A \cdot \sin\left(\frac{2\pi t}{T} + \alpha\right). \tag{2.2}$$

Expression (2.2) may seem somewhat odd, since the original equation does not contain the quantities A, T and α. Where do they come from and what do they represent? They originate from the requirement that $x(t)$ in equation 2.2 denotes the most general solution of the equation of motion (2.1). As such, $x(t)$ needs to be able to express the many ways in which an oscillator can be set into motion. In particular, parameters A and α take on that job and a graph of $x(t)$ can illustrate how. Figure 2.6 reproduces, for the specific choices of $\alpha = 0$, the possible values that the displacement from equilibrium may take on as time goes by. This graph is in the same spirit as the representation of the bead-in-a-box motion of figure 2.4. There, as well as in the present figure, for each moment *t* the value of the excursion *x* away from the equilibrium point is plotted. As the graph shows, an object whose position as a function of time is described by equation (2.2) remains inside a spatial interval ranging from $+A$ to $-A$, where A is the *amplitude* of the oscillation. The oscillator reaches these two extreme

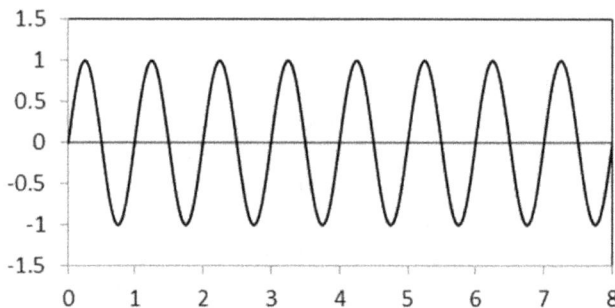

Figure 2.6. Position (vertical axis in units of the amplitude *A*) versus time (horizontal axis in units of the period *T*) trace for eight periods of a harmonic oscillator.

positions at moments in time separated by the interval T, the *period* of the motion. Of course, this is nothing but the periodic behavior of the sine function. The value in the brackets of the sine function in equation (2.2) determines the exact position of the oscillator. Thus, the argument of the sine function in equation (2.2) is called the *phase*. Just as the lunar phase tells us whether the Moon is waxing or waning, full or new, the phase of the oscillator informs us about its location: left or right, in transit through the zero point going left or right, etc. Note that, contrary to the particle-in-the-box case, the range of motion of the harmonic oscillator is not fixed. The stronger the initial kick, the larger the resulting amplitude.

The parameter α in equation (2.2) determines the displacement value at time $t = 0$. For example, if $\alpha = 0$, then, at time zero, the oscillator is at the origin ($x = 0$) as shown in figure 2.6. Similarly, if $\alpha = \pi/2$ the starting position is at the maximum possible ($x = A$). From a practical point of view, all it means is how we set the oscillator in motion at time $t = 0$. One straightforward approach is the following procedure. Take the mass between your fingers and gently pull it to a position of your choice, say $x = +A$. Then, at time $t = 0$, let it go without any further push or pull. In that scenario, the initial position is at $+A$ and the initial speed is zero speed. The ensuing motion is represented by the second example above ($\alpha = \pi/2$). Alternatively, at $t = 0$ we could impart a kick on the mass so that the initial speed is non-zero, while at that same instance the object is situated at $x = 0$ or any other point. Because the value of α determines the position of the oscillator at the starting moment, α is referred to as the initial phase of the oscillator. The behavior just discussed is reminiscent of the particle in the box, to which we can impart any velocity and which may initially be located anywhere inside the box. These types of parameters (here α and A) characterize the *initial condition* of the system under investigation, i.e. the position and velocity at time $t = 0$. Specification of these parameters is tantamount to specifying the energy that resides in the motion of the system. In general, the higher the energy, the larger the amplitude.

Harmonic oscillation exhibits a specific temporal correlation between position, velocity and acceleration. Figure 2.7 demonstrates this correlation for the case of a phase of $\pi/2$, i.e. an oscillator initially at rest and displaced to the right, i.e. $x(0) = +A$ and $v(0) = 0$. Immediately after the mass is released at $t = 0$, the stretched spring pulls the mass to the left. Thus its velocity becomes non-zero and negative which is consistent with picking up speed while going towards the left (positive velocity indicates motion to the right). At a quarter of the period, the speed reaches its maximum value (while still going to the left). At the same moment, the position is zero, indicating that the equilibrium is being crossed at full speed. After that, the magnitude of the velocity diminishes and becomes zero again at half the period, which is when the oscillator reaches the extreme of its displacement on the left ($x(T/2) = -A$). From there on, the motion turns to the right so that the velocity values are now increasing and positive. At $t = 3T/4$ the object passes through the equilibrium again ($x(3T/4) = 0$), now going from left to right, and reaches the right maximal position at $t = T$ when it momentarily comes to a halt again ($v(T)-0$). Just as the velocity describes the rate of change of position—$v(t)$ is proportional to the slope of $x(t)$—the acceleration is a measure of the rate of change of the velocity. The bottom

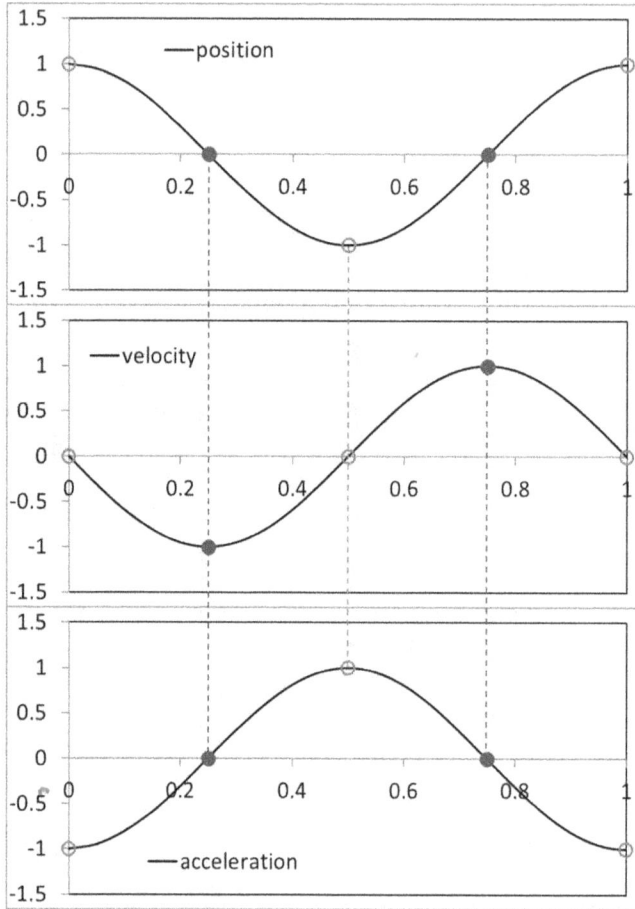

Figure 2.7. Position (top), velocity (middle) and acceleration (bottom) during one period of a harmonic oscillator with initial phase $\alpha = \pi/2$. Quantities are plotted in units of their respective maximum value.

panel shows that at any moment acceleration is directly proportional to the slope of the velocity in the middle panel. Maybe a language couched in terms of the acting force is more intuitive. No matter where exactly the object is, the spring force pulls the object back to the equilibrium point. If the mass moves away from the equilibrium, then this restoring force slows down the motion, and if the instantaneous motion is already towards the center point, then the force increases the instantaneous speed. That speed increase continues until the object crosses the midpoint $x = 0$ at maximum speed. Afterwards, the spring pulls the object back, i.e. slows it down until it reaches the opposite end, comes to an instantaneous stop, and the cycle repeats. Comparison of the top and bottom panels clearly shows that at any moment the position $x(t)$ and acceleration $a(t)$ are proportional to each other, except that if one is at a maximum, the other is at a minimum, and vice versa. In other words, the figure shows the validity of equation (2.1) in graphical form.

As long as they are realistic and do not lead to destructive behavior, the values of parameters associated with the initial condition are arbitrary. This is not true for the parameter T, the period of the motion. With the help of a dimensional analysis, we can gain insight into the dependence of the period T on the parameters k and m that describe the physical make-up of the oscillator. As long as the position of the oscillator is not zero, it follows from relation (2.1) that the ratio of acceleration a to position x has always the same value k/m. Because acceleration is the rate at which velocity changes with time and likewise velocity measures the rate of position change with time, the ratio a/x has the dimension of squared inverse time. The only characteristic time that enters into the analysis is the period T of the oscillator. Therefore, the ratio a/x and with it the constant k/m is inversely proportional to the square of the period T or in mathematical language: $k/m = c/T^2$ where c is a numerical factor that a dimensional argument cannot establish. A rigorous treatment reveals that a simple harmonic oscillator fulfills the following relation

$$\frac{1}{T^2} = \frac{1}{4\pi^2} \cdot \frac{k}{m} \quad \text{or} \quad T = 2\pi \cdot \sqrt{\frac{m}{k}}. \tag{2.3}$$

This period T is the intrinsic unit of time ticking away in any clock built on the simple harmonic oscillator—an important, non-trivial result. First, and most importantly, the period is entirely determined by the physical properties of the oscillator and does *not* depend on amplitude or starting conditions. This is the already proclaimed special property of linear spring motion. Secondly, equation (2.3) connects the period to properties of the spring that can be determined by static experiments: use a balance to find the mass m and measure the spring constant k via the extension due to an attached weight. Knowledge of these two parameters allows us to predict the oscillatory property of the mass–spring system. The mathematical relation (2.3) is not so simple that it could have been guessed based on physical intuition alone[10]. Sometimes, we do need mathematics to make progress. In any case, we now have a design tool at our disposal with which we can tailor the period of an oscillator to our needs. Usually, the spring is what it is and the value of the spring constant k is a given quantity. If we desire a specific value of the period T, say one second, then all we have to do is to attach an object with mass $m = kT^2/4\pi^2$, and voilà the clock ticks at intervals of one second.

2.6 Friction II: spoiling the simple harmonic oscillator

We are now in the position to discuss the influence of friction on the behavior of a harmonic oscillator. As it did in the particle-in-the-box clock, damping leads to loss of energy and thus a decrease of both the peak and average velocity of the oscillator. The details depend on the exact rate at which the oscillator loses energy. In general, the frictional force opposes the instantaneous motion and increases with increasing

---

[10] No matter how exactly dimensional analysis—the method of unit comparison used in the text above—is applied, the numerical factor 2π remains unknown.

speed. Already the simplest possible relation, that of a linear increase of the damping force with increasing speed leads to a model that has realistic features, with the added benefit of being mathematically tractable. As always, the ratio of the net sum of all forces and the mass of the object determine its acceleration. For a damped oscillator, Newton's equation of motion reads as follows:

$$F = -k \cdot x - b \cdot v \quad \text{and} \quad a = \frac{F}{m} \quad \Rightarrow \quad \frac{d^2x}{dt^2} = -\frac{k}{m} \cdot x - \frac{b}{m} \cdot \frac{dx}{dt}, \quad (2.4)$$

where the new parameter b characterizes the strength of the damping force and all of the other quantities have been introduced above. Note that I have used in the last part of equation (2.4) the explicit definition of acceleration a as being the rate of change of the rate of change of position (the rate of change of velocity), symbolized by the second derivative d^2x/dt^2. If we set $b = 0$ we recover the previous case of undamped motion. This differential equation is no longer solved by a simple harmonic oscillation, such as expression (2.2). In fact, three different solution types exist, depending on the relative strength of damping versus spring force. For us, the important scenario is that of 'weak damping', i.e. on average the damping force is much smaller than the spring force. The average damping force is $b \cdot \langle v \rangle$, where $\langle v \rangle$ is the average speed during one cycle. Likewise, the magnitude of the average spring force is $k \cdot \langle |x| \rangle$, where $\langle |x| \rangle$ is the average absolute displacement (ignoring directional distinction). The condition for weak damping is the requirement that, on average, the damping force is much smaller than the spring force which can be written succinctly as

$$b\langle v \rangle \ll k \cdot \langle |x| \rangle. \quad (2.5)$$

We can apply an order of magnitude analysis of this condition in order to gain insight into the relative magnitudes of the various physical parameters b, k and m. Blissfully disregarding the possible numerical influences of factors such as 2 or π (and even $4\pi^2$), we estimate that $\langle v \rangle \sim A/T$, where A and T are the amplitude and period of the oscillation, respectively. In the same vein, we replace $\langle |x| \rangle$ by A and then finally set $T^2 \sim m/k$ or $k \sim m/T^2$ (see equation (2.3)). Putting things together, we find that

$$\left(b\langle v \rangle \sim b\frac{A}{T} \right) \ll \left(k\langle |x| \rangle \sim kA \sim \frac{m}{T^2}A \right) \quad \Rightarrow \quad T \ll \frac{m}{b} = \tau. \quad (2.6)$$

The ratio of the mass m to the damping force constant b has units of time[11], so that we can compare the duration of the period T to the time span defined by m/b, for which we introduce the symbol τ. As can be seen from its definition, the characteristic time τ decreases with increasing damping force constant b. Increasing the damping force should lead to a faster decay of the oscillator's amplitude, and to a swifter decrease of its motion. Therefore, it seems reasonable to expect that τ is a

---

[11] Obviously $b \cdot v$ has units of a force and thus of *mass* times *acceleration* ($m \cdot a$). Therefore, the ratio b/m has units of *acceleration* divided by *speed* (a/v). In addition, because acceleration is the rate of speed change, it follows that b/m indeed has units of seconds and thus represents a time span.

measure for the time it takes for the average speed and average amplitude of the oscillator to diminish to a certain fraction of their initial values. The above line of argument is an example of how one can transform one equation (here: (2.5)) related to one feature (here: forces) into another equation (here: (2.6)) addressing a complementary aspect (here: characteristic times). Both relations are equally valid, so one can pick either one depending on circumstance or taste. And if nothing else, one point of view may initially be more intuitive than the other. While the outline here is somewhat superficial, it is borne out by a rigorous quantitative analysis of the motion of a damped oscillator characterized by equation (2.4). Such an analysis finds that for weak damping, the position of the oscillator is mathematically represented by the function

$$x(t) = A \cdot e^{-t/\tau} \cdot \sin\left(\frac{2\pi t}{T} + \alpha\right), \tag{2.7}$$

where all of the quantities have had their meaning introduced in the text above. The relation defined by equation (2.7) is a product of two functions—the exponential and sine functions. The latter describes the oscillatory aspect of the motion, and the former its unidirectional decay. The significance of this functional form will become clearer below. First, though, in an exactly analogous way to the particle-in-the-box, we can define the quality factor Q for the harmonic oscillator as the ratio of the damping time τ to the period T, i.e. $Q = \tau/T$. In particular, we will see that, for the damped harmonic oscillator, a relation between Q and energy loss exists that is similar to that of the freely moving bead. However, there are two major differences: (1) damping of the harmonic oscillator does *not* result in a changing period, as it does for the bouncing bead; (2) damping of the harmonic oscillator leads to a decrease of the amplitude, again unlike the bouncing bead. Figure 2.8 is the

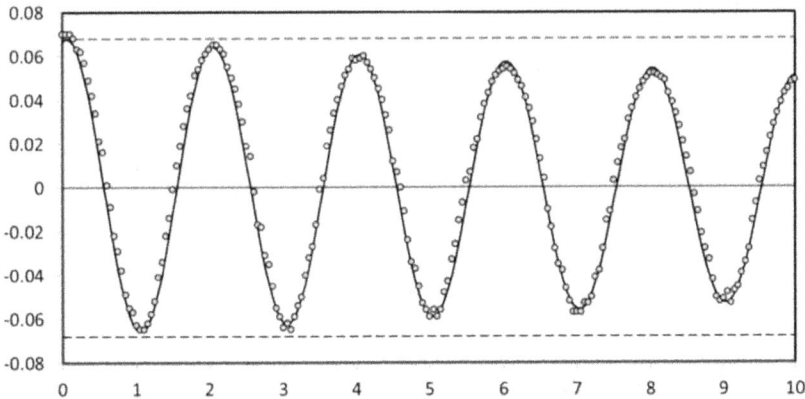

Figure 2.8. Decay of harmonic oscillator motion due to damping (aka friction). The initial amplitude, indicated by the dashed horizontal lines, decreases continuously over the five periods shown. A theoretical curve (solid line) of the form (2.7) is fitted to the measured values (open circles) of the position (vertical axis in units of meter) versus time (horizontal axis in units of seconds) of a mass gliding on an air track and attached to two springs fixed to the two ends of the track.

composite of a measurement performed on a real oscillator during one of my classes at Wesleyan (data represented by open circles) and the theoretical predicted curve given by equation 2.7 (solid line). For the latter the parameters A (amplitude of the oscillator), T (period), τ (decay time constant) and α (initial phase) have been adjusted to yield the best match with the data. It is not a perfect concurrence, but you will hopefully agree that the theoretical curve does a pretty good job in replicating the empirically observed behavior. We can therefore use equation 2.7 to find out how quickly (or slowly) the amplitude of a damped oscillator diminishes. For all practical purposes, the instantaneous amplitude for weak damping is determined by the exponential function at those times when the sine function is at a maximum. We will now see that, in contrast to the decreasing amplitude, the period of the damped oscillator is constant. The period of any damped oscillator represented by equation (2.7) is most readily observed as the time interval between two equivalent zero crossings, i.e. subsequent moments at which the oscillator moves in a given direction through the equilibrium point. From the properties of the sine function and the fact that the exponential function is never zero, those points in time are identified by the condition

$$\sin\left(\frac{2\pi t_n}{T} + \alpha\right) = 0 \quad \Rightarrow \quad \frac{2\pi t_n}{T} + \alpha = n \cdot \pi \quad \Rightarrow \quad t_n = n\frac{T}{2}, \qquad (2.8)$$

where the moments of zero crossing have been labeled t_n, and where n is any integer number 0, ±1, ±2, etc. Two subsequent zero crossings at t_n and t_{n+1} are exactly separated by an interval of $1/2T$, regardless of the specific value of n. In other words, no matter how long you wait and how much the *amplitude* has decayed, the *period* stays the same. At first blush, it therefore appears that, despite the presence of friction, we have hit on the perfect clock. Well, the oscillator is indeed a promising model, but we still have to solve the issue of how to overcome the decay without introducing irregularities into the motion of the oscillator.

2.7 Toy clock III: the pendulum

In the summer of 2015, the *Guinness Book of Records* added a somewhat unusual item. A pendulum clock—based on an old design, but only recently manufactured for the first time—was found to be working more precisely than any other pendulum clock swinging in air, although not any better than its inventor John Harrison (1693–1776) had claimed 250 years before. He had already proven his immense skills with a precision clock capable of determining longitude with unprecedented accuracy. Still, when he pronounced that his new clock would lose less than one second in 100 days, he was laughed at and worse. As an article in the 18 April 2015 edition of *The Guardian* newspaper reports [6], his announcement was judged at the time to be 'an incoherence and absurdity that was little short of the symptoms of insanity'. Apparently, Harrison's detractors who had haunted him earlier were still alive and well. It took over two centuries to vindicate him: in 2015, a clock based on Harrison's plans was measured—under the watchful eyes of the National Physics

Laboratory of the UK—to have lost only about 5/8 of one second in the time span of 100 days. Harrison's design was an achievement that was truly ahead of its time.

Pendulum clocks are part of the proud and long lineage of mechanical clocks, going back many centuries. While the story of mechanical clocks using foliot and verge, the oldest known type of clock escapement, has a hazy origin, the beginnings of the pendulum clock are somewhat clearer. At the start of the 17th century, Galileo had begun to investigate the motion of swinging objects. One specific aspect of this type of motion, namely its approximate independence of the duration of one swing of the length of the swing (the so-called isochronism), led him, in 1637, to the idea of a pendulum-based clock. But despite many attempts, neither he nor his son were able to construct such a clock. In 1656, Christiaan Huygens accomplished this feat by designing the first workable pendulum clock—and revolutionized time-keeping in the process. Given how simple a pendulum can be—tie a string around a stone and you have constructed one—the claim of being revolutionary may sound a bit like hyperbole. It is not. The first pendulum clock, manufactured for Huygens by the clockmaker Salomon Coster, improved the accuracy from about 15 minutes per day, typical for clocks then available, to about 15 seconds a day. In addition, the design allowed verge and foliot clocks to be retrofitted so that the regularity of the pendulum feedback soon improved the performance of many clocks in a variety of applications, not least astronomical observations.

The physics behind a pendulum is quite different from that of the harmonic oscillator described above. Physicists have so far identified four fundamental forces in nature, with gravity and electro-magnetic forces being the two most closely related to everyday experience. The pendulum needs gravity. It would not work onboard a free-falling spaceship such as the International Space Station. By contrast, the elasticity of the spring at the heart of the harmonic oscillator is a manifestation of electric forces[12] and does not require gravity at all. Despite this difference in the physics, the behavior of a pendulum and that of a harmonic oscillator are very similar. More precisely, for small amplitudes of the swing, the mathematical description of the motion of the pendulum is very nearly the same as that of the harmonic oscillator. As we have seen above, oscillatory motion comes about whenever the displacement from a point of zero force—the equilibrium point— gives rise to a force that pushes (or pulls) the object back with a strength that grows in proportion to the displacement from equilibrium. With a bit of geometry, we can show that the pendulum fits that bill. Once removed from its vertical position, the pendulum will oscillate with a period that depends solely on the length of the pendulum and the local strength of Earth's gravity, usually expressed by the gravitational acceleration near sea level, $g \approx 9.8$ m s^{-2}. However, in contrast to a mass attached to a spring, a mass attached to a string will swing with a period that does depend on the size of the swing.

To start, we inspect the instantaneous force acting on the bob of a pendulum, both when it hangs still in a vertical position and when it swings and makes a non-

---

[12] When a spring is stretched, the distances between the atoms of the spring increase ever so slightly. The electric charges that make up atoms (see chapter 4) respond with an increase in their mutual attraction. The macroscopic elastic force of the stretched spring is nothing but the sum of all these atomic scale electric forces.

zero angle with the vertical. According to Newton's equation of motion, objects at rest remain at rest if no force acts on them. Hence, it must be the case that in the vertical position no force acts on the pendulum bob. But of course, gravity has not been switched off—it still pulls the bob with its mass m downward with a strength mg. It is the string (or rod) to which the bob is attached that prevents it from falling. The bob pulls the lower end of the string down, creating tension along its axial direction, which exerts a comparable force on the upper suspension point. There, the tension of the string pulls the atoms of the material making up the support structure and this creates the necessary counter force so that the bob does not fall and remains at rest. It is complicated, but we will sweep the details back under the rug and just say that gravity and string tension cancel each other. Zero net force means zero acceleration and hence constant velocity. If the bob was at rest, it remains at rest. If the pendulum bob is laterally displaced, only the component of the gravitational force parallel to the string will cause tension, and the remaining portion—perpendicular to the string—is unbalanced and causes acceleration of the bob along the tangential direction, i.e. along the arc of the swing. Geometry dictates that these two components are related to the angle between the instantaneous direction of the string and the vertical. Calling the angle θ (the Greek alphabet is a source of useful and beautiful symbols), the tension causing component is given by $mg \cdot \cos \theta$, while the accelerating part is given by $mg \cdot \sin \theta$. This is an ominous development. If we are to find the motion of the pendulum to be similar to that of the harmonic oscillator, we probably cannot afford for the force to be anything but proportional to the angle θ. The sin function will spoil that. Indeed, the linear acceleration a along the tangential direction is proportional to the angular acceleration associated with the angle θ, as well as the length L of the pendulum. Therefore, analogously to the equation of motion of the harmonic oscillator, the dynamics of the pendulum is described by

$$ma = mL\frac{d^2\theta}{dt^2} = -mg \cdot \sin \theta \quad \Rightarrow \quad \frac{d^2\theta}{dt^2} = -\frac{g}{L} \cdot \sin \theta, \qquad (2.9)$$

which is definitely not of the form of the harmonic oscillator. Fortunately, equation (2.9) still has solutions that are repetitive, with a fixed period T. Unfortunately, however, this period depends, in a complicated way, on the amplitude of the swing! Out the window goes Galileo's isochronism. What happened? Was he mistaken? Yes and no. Strictly speaking, the pendulum period changes with amplitude. In this regard, the pendulum behaves like the bouncing bead. However, as long as the maximum angle of the pendulum is 'sufficiently small' (more on that in a moment), the amplitude dependence becomes weak and in this limit the sin function can be replaced by its argument θ. With this approximation, equation (2.9) assumes a form that is mathematically equivalent to that of the harmonic oscillator: replace θ by x, g by k and L by m, and you turn (2.9) into equation (2.1) of the harmonic oscillator. Using the reverse substitution, equation (2.3) for the period T of the harmonic oscillator turns into the period of the pendulum, i.e.

$$T = 2\pi \cdot \sqrt{\frac{L}{g}}. \qquad (2.10)$$

By sheer cosmic coincidence, the period of a one meter-long pendulum at sea level is only a few milliseconds longer than two seconds (more precisely 2.006 s). Before the French Revolution, much thought was given to the proposal to use the second pendulum, as it was called, to define the unit *length*. However, because the exact value of the gravitational acceleration depends on latitude and height above sea level, the proposal was eventually abandoned. We will see in chapter 4 that the time indicated by atomic clocks—and for that matter by *any* clock—also depends on its vertical position. That, however, turns out to be related to the entirely different story of general relativity.

So, how small should the angle θ be for equations (2.9) and (2.10) to be valid? As always, that depends on how much approximation you are willing to tolerate. The question concerning the proper equation of motion is easier to answer. Up to angles of about 15°, the error introduced by the linearization of the sin function is not more than 1%. For the period, the situation is a bit more complicated. Suffice it to say that the *exact* period of the pendulum, not the *approximate* expression (2.10), can be written as an infinite (yes, infinite) sum. For most purposes, specifying the three leading terms is sufficient:

$$T = 2\pi \cdot \sqrt{\frac{L}{g}} \left(1 + \frac{1}{16}\theta^2 + \frac{11}{3072}\theta^4 + \cdots \right).$$

To lowest order, the difference between the exact period and the approximate period is proportional to the square of the angular displacement θ. Therefore the error from making the approximation rises quickly. The percentage difference reaches 1% at an angle of about 13° and changes by ±0.2% for an amplitude as little as ±1°. Early pendulum clocks relied on wide swings, so that even relatively small amplitude changes due to friction caused fairly large period variations. One of the improvements of Huygens' design was the restriction to small amplitudes. If the amplitude is reduced to 5° or less, then a relative amplitude variation of one part in ten, as used above, causes a period variation of not more than about ±0.015% or one millisecond for the meter pendulum. After one day, the accumulated uncertainty grows to about 15 seconds. In the 17th century, you certainly could live with such ambiguity—unless you were interested in measuring longitude.

2.8 Resonance and feedback

The soprano's voice moves up and down the scale only to home in on one single pitch. Obviously the right one since the audience is visibly moved. Then, with increasing volume and another slight adjustment of the pitch, the audience can no longer contain itself and bursts into ... pieces. Being a wine glass of a certain shape and thickness it had no choice. Such a feat of shattering a glass with your vocal cords can indeed be accomplished [7] albeit only if you command over extreme lung power—or a good loudspeaker and amplifier.

When acoustic waves push and pull at a specific rate, the glass—or any other object for that matter—is forced to oscillate at the same rhythm as the audio stimulus. Wine glasses and bells have similar shapes, so it is hardly surprising that they ring with specific tones that depend on form and make-up. But not just bells, *all* objects have more or less well defined natural frequencies at which they can shudder and tremble most effectively. If the rate of external pulling and pushing coincides with one (or more) of these intrinsic frequencies, resonance is established. Since all of us—I hope—have the experience of being on a swing, getting higher and higher with just the right body moves by either yourself or your friend pushing the swing, we have an intuitive feel for what resonance is and does. But as scientists, we want a deeper and more detailed understanding. Furthermore, in the current context of clocks, resonance plays a specifically important role. So let us investigate this phenomenon a bit further, bearing in mind that the chosen *specific* example illustrates a *general* behavior.

In the case of a swing it is clear that proper timing of the external forcing is crucial. However, the exact nature of response and stimulus is very likely obscured because whoever sits on the swing will not sit still. Part of the fun is to move and to stretch out those legs. Replace the person on the swing (sorry) with a weight and you have a pendulum. Set into motion by an outside force, the amplitude of motion of a given swing depends only on the properties of the external agent. We would expect the strength of the force and the rate at which it changes to be relevant. In the simplest scenario these two quantities are also sufficient to describe the stimulus part of the system both qualitatively and quantitatively. How about the response part? The resulting amplitude will certainly be important, but there is another, maybe less obvious aspect. Rome was not built in one day and, in general, you will not be able to get the swing to its maximum excursion in one single push. Only after a certain 'build-up' time, will the system reach steady-state and swing with constant amplitude[13]. Likewise, the free pendulum needs some time to come to rest once the external stimulus has ended. The two time scales are related to the degree of friction present—the less of it there is, the longer both build-up and decay times are. Friction has another, less obvious influence that is best elucidated by describing an experiment one might perform in order to quantify the resonance behavior of any oscillating system subject to a repetitive driving force. Here is what you should do: apply the repetitive force with known (measured) strength and frequency. Wait an appropriate length of time so that the steady state condition is reached, at least to the degree of precision you care about. Then measure the amplitude of the driven oscillator. Paying close attention—as all good scientists can be expected to do—you notice that force and oscillator do not attain their respective maximum at the same time. Rather, depending on frequency, the oscillator leads or lags the force, which is important to know if you are interested in the energy exchange between driving agent and driven object. However, this detail is not our concern here. When you perform experiments, several parameters can usually influence the results. Some of those are more or less given, for example the gravitational pull by Earth on a pendulum, but others you can tweak. First, we must settle on a specific oscillator and

---

[13] If you want to see an impressive example of such a build-up, just search for 'botafumeiro' on the web and have a look at one of the many videos showing how, in the church of Santiago de Compostela in Spain, a vessel with burning incense is being coaxed to swing with eventually enormous amplitude.

a specific mechanism to couple the external force. Then, next to force strength and rate, the overall friction or damping of the system plays the most important role for the outcome of our measurements. In order to quantify the drag on the oscillator, we can observe the waning of the amplitude of the free, unforced oscillator—any characteristic decay time, call it τ, will do, say the time it takes to cut the initial amplitude in half. Thus we have four quantities that we need to measure: strength F_0 of the external force, its period T or equivalently its frequency $f = 1/T$, the decay time τ and the steady-state amplitude A at which the driven oscillator (or pendulum or whatever) moves. As a reminder, in steady state, the forced oscillation occurs at exactly the same frequency f at which the system is driven, i.e. the frequency of the oscillator is *not* another variable we would have to keep track of. Amazingly, the results you will find can be summarized in a single graph that is valid for almost any type of oscillator and driving agent—whether pendulum, mass on a spring, atoms driven by electromagnetic radiation or certain electric circuits made of wire, coils, and capacitors.

By way of example, figure 2.9 shows how the steady state amplitude (vertical axis) of a simple oscillator with one natural frequency, here of 40 Hz, changes when it is driven by an external force with adjustable frequency f (horizontal axis). While the figure shows *calculated* results, it is a realistic representation of what you would find in a series of experiments, carefully executed in the spirit outlined above. The solid curves in the figure represent two cases of different friction ($Q = 5$ and 20, see section 2.4 for a definition of the quality factor Q), but otherwise identical properties—the red/black line (large/small peak) going along with the case of low/high friction, respectively. In order to facilitate comparison of the *widths* of the two curves, the high friction case has been scaled so as to exhibit the same maximum and is shown as the dashed curve. In other words, solid and dashed black curves differ only by a multiplicative constant. Figure 2.9 contains a wealth of information on the general resonance phenomenon. It also indicates what properties we need for a precise,

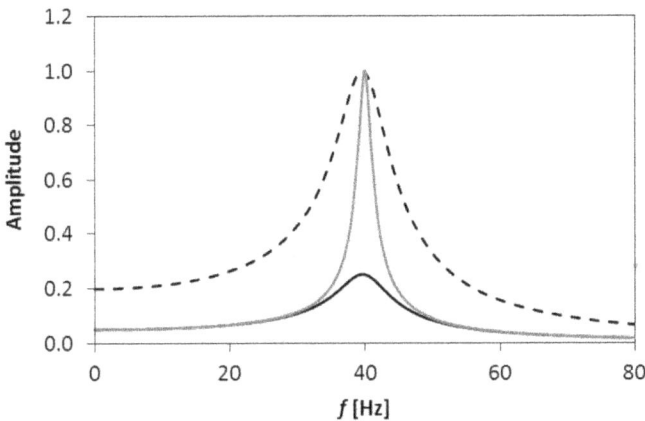

Figure 2.9. Resonance behavior of a driven oscillator with natural frequency 40 Hz and Q-factors of 5 (solid black curve) and 20 (solid red curve). The dashed line is a scaled version of the high friction case $Q = 5$.

accurate clock, as a brief discussion of the salient features of figure 2.9 will demonstrate next.

When the rate of the external force near-matches the natural frequency of the oscillator, the latter responds with the largest amplitude. This is clearly the message of figure 2.9 for both cases of damping and this is of course the very nature of resonance. However, damping has an important influence as it washes out the resonance in two ways. First, it obviously diminishes the amplitude maximum that occurs at resonance. The reduction factor for the present example is close to the numerical value 4, the ratio of the Q factors of our two cases. Theory does indeed predict that the resonance amplitude increases in approximate proportion to Q so that enormous amplification can be achieved in the limit of large quality factors. Furthermore, increasing damping of the oscillator results in a broadening of the resonance curve as the lines in figure 2.9 are called. This aspect in particular spoils the selectivity of resonance. For the case of $Q = 20$, the low damping case, the driving frequency has to be close to its resonance value in order to achieve amplitudes larger than say half the maximum. For the higher damping scenario of $Q = 5$, a significantly larger range of frequencies are permissible to achieve the same goal. It is this specific feature of generic resonance behavior that is important for clocks.

Suppose you want to build a pendulum clock. Or a quartz crystal clock whose time piece more resembles a harmonic oscillator. Or an atomic clock whose hardware does not resemble any of the above, but whose properties can be understood and modelled in exactly the terms discussed in this chapter (see section 3.5 'Inner workings and limits of atomic clocks', and figure 3.7 in particular). In any case, you are concerned that the clock's time keeper should oscillate at a rate that is as uniform as technically possible. Because of unavoidable friction, clocks must be kept alive by an external driving mechanism whose own frequency is usually not as reliable as that of your clock. Since a driven oscillator adopts the frequency of the driver, how can you guard against this unwelcome influence? Enter feedback. As long as you can sense the instantaneous amplitude of a driven oscillator, you can also sense if the amplitude changes. When the driver frequency drifts away from the optimum, i.e. the resonance frequency, the amplitude responds in only one way—it decreases. Amplitude decrease also occurs when the driver frequency is off-resonance and roams even further away from the optimum (see figure 2.9). However, if the frequency returns to its resonance value, the amplitude will increase until the maximum is reached. Therefore, it is possible, by proper assessment of measured *amplitude* values, to keep the driving *frequency* within a narrow range around its uniquely determined resonance value. The sharper the peak at resonance is, the smaller the frequency interval is in which the clock time keeper can drift. All high precision clocks, from pendulum to quartz to atomic clock, rely on sophisticated feedback mechanisms to keep the clock frequency from straying.

2.9 Numbers for comparison

Concepts are important in physics and so are numbers—we have seen that already in other contexts. Here we want to acquire an understanding of 'typical' values for the various parameters involved in the specification of the period of an oscillator, equation (2.3), or a pendulum, equation (2.10). For starters, what sorts of periods can we expect for a mass on a spring? Based on relation (2.3), there is no limit in principle for either really long or really short periods. While this chapter is concerned with small time units, mechanical oscillators are also capable of marking off long time intervals. A fascinating project is currently underway to construct a 'millennium clock' designed to run for 10 000 years. A first prototype [8], which began to tick on 31 December 01999 (note the leading zero: the Millennium Project uses, quite consistently, five digits to designate years), uses a torsional oscillator as its timekeeper. It is not a mass-on-a-spring-type mechanism, but the restoring force of a twisted wire leads to angular oscillations that are described in an entirely analogous fashion to equations (2.1) and (2.2).

On the short duration side, mechanical oscillators in the form of small quartz crystals can vibrate with periods that are a small fraction, actually about 1/30, of a millisecond. Usually, the frequency of such crystals commonly used in watches is $f = 32\,768$ Hz with a corresponding period of $T = 1/f \approx 0.0305$ ms. One can view the quartz crystal as an assembly of a colossal number of miniature mass and spring oscillators. Molecules making up the crystal are interacting with their neighbors in such a way that each one of them is held to its equilibrium position, around which it can vibrate very much like a mass on a spring. Consequently, the entire crystal expands and contracts, albeit with a tiny amplitude. Making the size of the crystal small—which is anyway needed to fit it into a watch—ensures that the mass entering into formula (2.3) is low. On the other hand, the strength of the molecular interaction and therefore the effective spring constant k is very large. Both aspects together explain the remarkably short oscillation period of the crystal. The inner workings and a bit of history of quartz clocks will be one topic in the next chapter.

Periods of 'ordinary' springs and masses range from fractions of a second to seconds. Take, for example, a fairly soft spring that stretches by 20 cm when a mass of one kilogram is attached. The restoring force of the spring $(-k \cdot x)$ is then in precise balance with the gravitational pull on the mass $(-m \cdot g)$ and $k \cdot x = m \cdot g$, with $g = 9.8\,\mathrm{ms}^{-2}$. Thus the spring constant follows as $k = m \cdot g / x \approx 9.8 \cdot 1/0.2\,\mathrm{kg \cdot s}^{-2} \approx 49\,\mathrm{N\,m}^{-1}$. If that system is now perturbed and the mass is pulled down or pushed up slightly, the spring force is either larger (pulled down) or smaller (pushed up) than the constant weight $m \cdot g$ causing oscillation around the equilibrium point with a period of almost one second:

$$T = 2\pi \cdot \sqrt{\frac{m}{k}} \approx 2\pi \cdot \sqrt{\frac{1}{49}}\ \mathrm{s} \approx 0.9\ \mathrm{s}.$$

Next let us inspect more realistic oscillators, those that are damped and whose amplitude will decay unless a power source provides feedback. We saw earlier in this chapter that the numerical value of the Q-factor of an oscillating system is roughly equivalent to the number or periods it can swing freely without the need for feedback. Thus a one-second pendulum clock could run for about two weeks or so without external input if you manage a Q-value of about one million. But there is no simple way to eliminate friction from any mechanical mass–spring system to such an extent—notwithstanding the fact that a massive pendulum, used as a gravity wave detector, features a Q-value of about 20 million [9]. That was only possible with a heroic effort. More realistically, Q factors of the order of $10^4 = 10\,000$ are possible which amounts to a free running time of about three hours. This is still quite a challenge.

Pendulum clocks do not lend themselves to measure short time spans. As we have seen above, the period of a pendulum scales as the square root of the ratio of the pendulum length to the local gravitational acceleration. Earth pulls everything down at a rate of about 9.8 m s^{-2} and even on Jupiter—the most massive planet in the solar system—the equivalent gravitational acceleration is 'only' about three times larger. According to equation (2.10), an earthly pendulum that is 10 mm long has a theoretical period of about a fifth of a second. I do not know whether anyone has tried, but manufacturing such a small pendulum would provide significant challenges—and would still not yield a clock capable of measuring particularly short time spans. The forte of the pendulum is the more leisurely one second plus pace. We saw above that we need a length of about one meter for that period. Increase the length and the pendulum swings longer. I suppose I should consult the *Guinness Book of Records* for the slowest pendulum ever built, but some of the so-called Foucault pendulums must rank up there. In the former church of Sainte Geneviève in Paris, now known as the Panthéon, you can still admire a copy of the pendulum Foucault hung up there for the Paris Exhibition in 1850. Its 67 meter length gives it a period of about 16.4 seconds. At the latitude of Paris, the plane of the pendulum rotates by about 11° in an hour. For an assumed swing amplitude of the pendulum of 2 m, the lateral displacement of the pendulum bob after 2 min is already about 1.3 cm—a noticeable change.

2.10 From Huygens to shortt—how the pendulum revolutionized time-keeping

When Galileo first thought of using the pendulum as a clock, mechanical clocks had been already around for several centuries. They were not very reliable, though, and were typically off true time—for example, as measured by the appearance of noon—by minutes after a single day. Therefore, the Sun was used to recalibrate the clock as needed. Astronomical applications in general, but also requirements associated with regulating life in monasteries and other institutions, provided ongoing incentives to look for improvements. More importantly, at Huygens' time, ocean-faring nations had discovered that you could find your way along the east–west direction by comparing directly observed local noon with the time indicated by a precise, reliable clock that was set to show noon at the port of departure. Every hour's difference implies that one has sailed 1/24 of Earth's circumference in the longitudinal direction. So if the clock can keep time to 15 seconds or 1/240 of an hour per day—as Huygens' pendulum clock could—then at the equator the east–west position is known to within 1/240×1/24 of the globe's circumference, or about 7 km. With journeys lasting many days or even weeks, this uncertainty adds up to a significant amount. Thus, much more precise clocks were highly desirable. No wonder, Harrison's proposed clock with a precision of 1 second per 100 days was greeted with strong reactions.

The highest precision ever achieved with a pendulum clock is quite a bit better than is possible even with Harrison's design. Based on electro-magnetic feedback, refinements of the pivot points, release mechanisms and other crucial parts, as well as careful minimization of friction—including that of air by letting the pendulum swing in an evacuated container—resulted in the record-breaking stability for the Shortt–Synchronome clock. During the 1930s and 1940s, this pendulum clock set the standard for precision time-keeping . A recent investigation of a specimen still kept at the US Naval Observatory revealed that the intrinsic precision of this clock is astonishing. According to this study, the instrument is capable of telling time with a precision of about 1/5 of a millisecond per day, which amounts to losing or gaining not more than a second in about 12 years. No other electro-mechanical clock comes close. The use of electromagnetic schemes to provide energy input overcomes the need to rewind the weights that previously powered all-mechanical clocks. First batteries, and then later AC electrical power, provided an essentially frictionless and continuous feed of energy into the pendulum motion.

As we saw at the start of this chapter, a clock deserving its name must have both a timekeeper and a means to indicate or register time. The linkage of these parts introduces friction and disturbances, even if the timekeeper is otherwise operating perfectly. Until the beginning of the 19th century, mechanical coupling was the only possible approach, although it unavoidably introduced fairly large amounts of friction. A common answer to that challenge was to supply lubricants for moving parts. However, lubricants collect dust and/or change in composition and become less efficient over time. Therefore, the best timepieces—including the Harrison clock and modern, expensive watches—dispense with oil and grease, opting instead for materials selected for smoothness, such as precious stones and high-quality steel for pivot points and rotation axes. No matter how sophisticated these choices, in the end residual friction limits the performance of purely mechanical clocks. Progress came in the form of using electrical impulses to control the necessary energy input and information output. Furthermore, isolating the timekeeper as much as possible from the rest of the clock proved essential. In the Shortt clock, a primary pendulum, swinging in vacuum and engineered to minimize friction at all cost, has the sole function of producing a constant interval of time. Electrical pulses communicate that rate of oscillation to a secondary pendulum, thus avoiding any mechanical contact. With the help of this feedback, the secondary pendulum is *forced* to stay in synchrony with the primary one, regardless of any interfering influence of the readout mechanisms applied to it. Likewise, contactless electrical pathways provide what little energy is required to keep the primary pendulum going. To be sure, the elements and circuits necessary to make it all work are non-trivial. But work it does.

The end of the 19th and the beginning of the 20th century saw a proliferation of devices that joined electro-magnetic and mechanical function. Time synchronization demands by railroad and telegraph companies were a powerful stimulus for this development. The Shortt clock was just the culmination in a long history of similar clocks and other related devices[14]. Many patents were filed for these inventions, some at a small office in Bern, Switzerland. We will see in chapter 4 what a clerk there had to contribute to the story of clocks and time. But first, we continue with the

---

[14] A brief account of the history of electromagnetic pendulum clocks can be found in [10], including animations of the Shortt pendulum clock and various other mechanisms.

next chapter in time-telling precision, in which vastly superior accuracy was made possible by an entirely new type of clock based on the atom, the tiny building blocks in nature's tinker toy construction set.

References

[1] http://members.aon.at/sundials/berlin-egypte.htm

[2] Alder K 2002 *The measure of all things: the seven-year odyssey and hidden error that transformed the world* (New York: Simon and Schuster)

[3] Dershowitz N and Reingold M 2008 *Calendrical calculations* (Cambridge: Cambridge University Press)

[4] Dohrn van Rossum G 1996 *History of the Hour* transl. T Dunlap (Chicago: Chicago University Press) p 239

[5] Sullivan D B 2001 Time and frequency measurement at NIST: The first 100 years Proc. of the 2001 *Int. Frequency Control Symp. and PDA Exhibition IEEE* pp 4–17 www.nist.gov/pml/div688/grp40/division-history.cfm

[6] McKie R 2015 Clockmaker John Harrison vindicated 250 years after 'absurd' claims *The Guardian* **18** April
www.theguardian.com/science/2015/apr/19/clockmaker-john-harrison-vindicated-250-years-absurd-claims

[7] Schrock K 2007 Fact or fiction? An opera singer's piercing voice can shatter glass *Sci. Am.* August 23 www.scientificamerican.com/article/fact-or-fiction-opera-singer-can-shatter-glass/

[8] http://longnow.org/clock/prototype1 (website of The Long Now Foundation)

[9] Cagnoli C *et al* 2000 Very high Q measurements on a fused silica monolithic pendulum for use in enhanced gravity wave detectors *Phys. Rev. Lett.* **85** 2442

[10] Bosschieter J E 2000 *Electric clocks—A history of the evolution of electric clocks* ed Federatie van Klokkenvrienden (The Netherlands: TIJDschrift) www.electric-clocks.nl/clocks/en/index.htm

Chapter 3

From milliseconds to attoseconds: is there a limit?

A picture is worth a thousand words, as the saying goes. It is probably true, since I need about half that number just to explain in part what figure 3.1 is trying to communicate[1]. To put it succinctly, the graphics shows how 700 years of tinkering with clocks has improved their time-keeping accuracy. The horizontal axis displays the approximate time, from the year 1300 to the first century of the current millennium, in which a type of clock was first introduced. Along the vertical axis, time-keeping accuracy, measured in 'lost seconds per day', progresses on a logarithmic scale, i.e. numerical values are multiplied or divided by a factor of 10 for each equidistant vertical step down or up, as marked off by the horizontal lines. Note that the order is inverted: the largest value (1000) is at the bottom and the smallest (one-trillionth) is at the top. From these two limiting numbers, it should be clear that to speak of 'improvement' in clock technology over the time span in question is an understatement. A change in performance of a trillion and more— much of that in the last 100 years—is more consistent with a revolution in time-keeping. This chapter will provide a closer look at the driving forces for this development and the new physics that made it all possible.

The first entry (A) in the graph refers to a 'typical' verge and foliot clock from the early 14th century. These timekeepers were off from 'true time' by 10 to 15 minutes in a single day and thus 'lost' about 1000 seconds per day. In less than a week, the clock would have announced noon an hour early or late. In comparison with modern standards, such performance might seem pitiful, but it served the contemporary need for precision. Progress in the design and manufacture of mechanical clocks did not result in major improvements (B) until Huygens' invention of the pendulum clock (C). Graham, Harrison and other gifted clock makers improved the

---

[1] Many versions of this graph exist in the literature. See, for example the NIST webpage [1].

doi:10.1088/978-1-6817-4096-6ch3   3-1

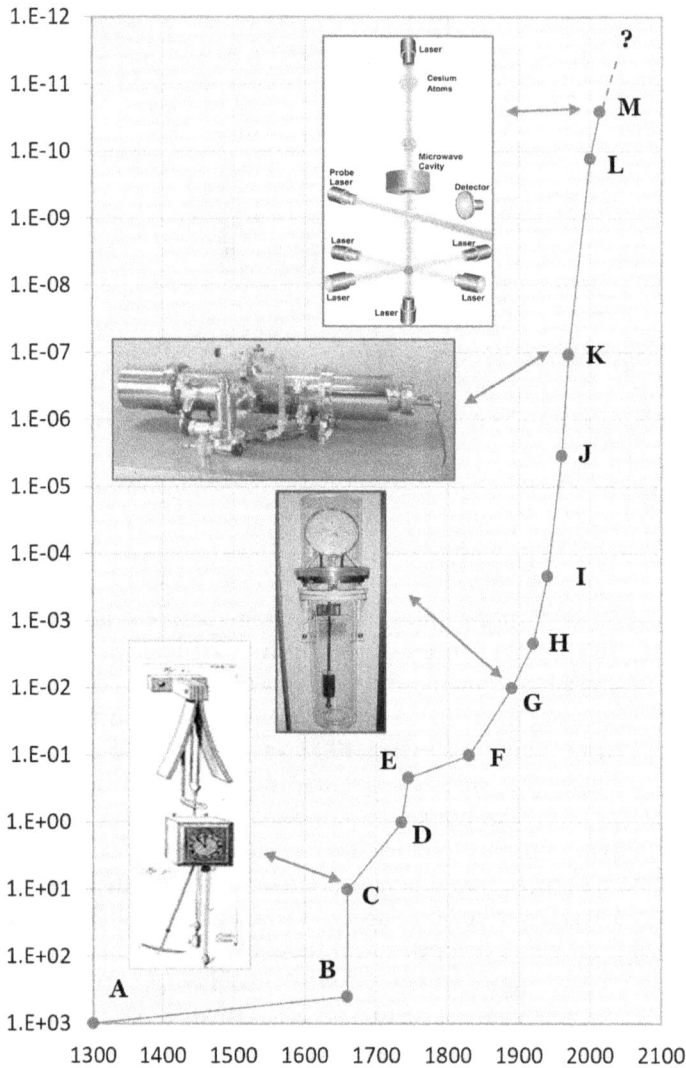

Figure 3.1. History of clock performance measured in 'lost seconds per day' is shown for various types of clocks against the year of first introduction (see text for meaning of labels).

precision and reproducibility of mechanical timepieces even further (D–F). This development culminated in the free pendulum clocks by Rieffer and Shortt, which kept an impressive regularity by a finely tuned feedback loop (G and H). They served as time standards in the early parts of the 20th century. A recent investigation [2] indicated that the accuracy of the Rieffer–Shortt pendulum clock might have been 0.2 ms per day, even better than is indicated in figure 3.1. If correct, this would rival the next entry, item (I), which denotes another significant stage in the history of clock making, namely the invention of the quartz clock. Without something fundamentally new, the story captured by the graph would probably end about there. Items J and following are indeed something completely new, they represent atomic clocks up to the current standard-bearer as the highest precision clock, the cesium fountain clock (M).

With the introduction of atomic clocks came a new definition of the unit of time. Our good old planet was found to be lacking in the precision and stability of its rotation. Like ice skaters who increase and decrease the swiftness of their pirouettes by moving their arms closer to or further away from their body, the seasonal movement of water up and down in the atmosphere imprints an annual rhythm onto Earth's rotational rate. Tidal forces from the Moon coupled with friction in the crust cause a secular slowing down. These and other, in part unexplained, irregularities make Earth's rotation (see also figure 3.12) less useful as a standard than the precise, reproducible and—as far as we currently know—immutable oscillations of atoms. Which particular atom in which specific state we use as a time standard is again a matter of convenience and our ability to use precision measurement tools. Since 1960 and until now (2017), the SI unit of one second is defined to be equal to the duration of 9 192 631 770 oscillations of cesium atoms when they are kept in a specific state of excitation (more on that below).

In the previous chapter, we discussed the main ingredients that make up a clock. In some ways, faulty time-keeping can come about through imperfections in any of the parts. If, for example, the energy input fails, the clock stops. And the best drive mechanism will be wasted if the counter changes value with random errors without feedback from the clock or if the indicator becomes stuck. However, optimizing these clock aspects requires engineering that is not relevant to our central topic. For the present purpose, we will concentrate on just one aspect, namely the 'smoothness', 'regularity', 'exactness' and 'reproducibility' of the oscillator. I have put the terms in quotation marks to indicate that we need a deeper understanding of this require-ment. To get a sense for the conundrum, consider the following. Clearly, to determine that two ticks of a given clock have the same length, we need another clock. But then how do we ascertain the 'exactness' of *that* clock? For the moment then let's be satisfied by stating the obvious: a good clock is one for which, each tick and each swing is the same as that before or after.

In everyday language, the words *accurate* and *precise* are often used interchange-ably. With relation to measurement, we have to pay more attention to detail. If we accept the premise that the quantity of interest has a true value, then the *accuracy* of a measured value is judged by its deviation from that true value, while the *precision* is determined by the number of digits with which the quantity can be specified. For example, the length of the sidereal day can be measured by the time between two subsequent meridian transits of the star Aries. Suppose this interval has been determined to be 86 163 s with an uncertainty of ±2 s. Then the precision of this measurement is two parts in 86 163. However, given that the otherwise known length of the sidereal day of 86 164.09 s, the accuracy of the observation is about one part in 86 164, twice as good as the precision. On the other hand, using a clock with a millisecond readout but faulty timing (say off by a minute a day) might yield a result of 86 102.093 s, which impresses with high but meaningless precision, since it fails quite badly in terms of accuracy.

Before we get started with the main part of the chapter, I want to clarify its title. Following the convention of the SI metric system for fractional units in general, successive division by a factor 1000 generates the prefixes and fractions of the temporal base unit of one second in table 3.1.

3.1 From quartz vibrations...

When you visit the Physics department at Wesleyan University and take the elevator to the second floor of the Exley Science Building (I prefer the stairs, but they are not as easy to find), right across the hallway is the Cady Lounge. This large room occupies a prominent place and serves as a central meeting place. If you ask students, they may know or guess that Cady was a former 'physics prof'. Probably only a few students will be able to tell you more. You might have slightly better luck asking a current faculty member. From 1902 to 1946, Walter Guyton Cady was indeed a physics professor at Wesleyan, where he discovered—among other things—that quartz crystals have sharp resonances. Which is a helpful thing to know when you want to build a good clock. But that was not the first thing on Cady's mind. He was more interested in radio technology.

Cady's father ran a business supplying provisions and warehouse storage to merchant vessels and his grandfather had been a sea captain and a West Indies merchant [6]. Oceangoing trade was not to be our Cady's thing, although in the late 1890s he did cross the Atlantic to pursue studies in Germany. After graduating from Brown University, he received a PhD in physics from the University of Berlin, and worked for a while with what is today the National Geodetic Survey before taking up a professorship at Wesleyan University. Cady's original research interest shifted from arc discharges and radio detectors to a phenomenon called piezoelectricity. By the second half of the 19th century and picking up speed during World War I, much of the technological progress in industry and private households came from innovations related to things electric (see figure 3.2 for a glimpse of that

Table 3.1 List of the units of time derived from the SI base unit of the second by subsequent division by 1000. Examples are completely arbitrary.

Time unit	Symbol and value	Example
Second	$1\text{ s} = 1000\text{ ms}$	An elephant heart needs about 2 s for one beat [3]
Millisecond	$1\text{ ms} = 10^{-3}\text{ s} = 1000\text{ μs}$	Approximate spin period of fastest known pulsar.
Microsecond	$1\text{ μs} = 10^{-6}\text{ s} = 1000\text{ ns}$	Clock cycle of an Intel microprocessor of the 1980s
Nanosecond	$1\text{ ns} = 10^{-9}\text{ s} = 1000\text{ ps}$	Exposure time of an intensified CCD camera
Picosecond	$1\text{ ps} = 10^{-12}\text{ s} = 1000\text{ fs}$	Period of molecular vibration
Femtosecond	$1\text{ fs} = 10^{-15}\text{ s} = 1000\text{ as}$	Period of UV light (300 nm)
Attosecond	$1\text{ as} = 10^{-18}\text{ s} = 1000\text{ zs}$	Currently shortest laser pulse, shortest time measured [4]
Zeptosecond	$1\text{ zs} = 10^{-21}\text{ s} = 1000\text{ ys}$	
Yoctosecond	$1\text{ ys} = 10^{-24}\text{ s}$	Light travels across a proton in about 3 ys [5]
Planck time	$5.4 \cdot 10^{-44}\text{ s}$	Possibly the briefest meaningful time span (see chapter 6)

Figure 3.2. Electrical power lines in New York City in the 1890s. Reproduced from *Book of Old New York* (Henry Collins Brown 1913).

advancement—curiously, some places still look like that): electro-motors, the telegraph, the telephone, lighting, radio and other new-fangled gadgets were all based on electro-magnetic processes and phenomena, helped by a relatively new and deep understanding of electricity and magnetism. This understanding had been coming for a long time and too many names, too many chapters are associated with its story to do it justice here. So, I just pick one name, that of James Clark Maxwell, who single-handedly put into succinct mathematical language the evidence, observations and deductions of hundreds of people, collected over hundreds of years. Four elegant and very general equations, now known as Maxwell's equations, augmented by relations that are valid for any specific substance that might be involved, not only summarize all electromagnetic experimentations that had been performed by Maxwell's time, but are also at the heart of those that were yet to come. Maxwell's equations pertain to electric and magnetic processes that play out in a vacuum, in the absence of material objects. The response of matter to electric and magnetic stimuli is coded in mathematical relations for pertinent properties of the specific material of interest, using either an empirical or a theoretical approach. An example for the former type is Ohm's law, which expresses the observation that, over a wide range of parameter values, the electric current sent through a material of given geometry is—at least approximately—proportional to the voltage applied to the opposite ends of the object. While this empirical relation can be derived from theoretical arguments, Maxwell's equations alone are insufficient for the job. In any case, the large diversity of substances we have at our disposal—metals, ceramics, glasses and organic substances to name a few broad categories—explains the enormous variety of electrical phenomena, among them the piezoelectric effect.

Take a cylindrically shaped piece of quartz[2]. Attach tightly fitting thin metal rings around each end of the cylinder. With the help of wires connect the two poles of a battery to the metal rings. As soon as you apply the battery voltage to the metal rings, the quartz cylinder changes length slightly. Reversing the polarity of the applied voltage results in alternating contraction and expansion of the cylinder. Conversely, change in length, i.e. mechanical stress of a quartz crystal, leads to a voltage between the opposite ends of the quartz crystal, as the Curie brothers had already been demonstrated in 1880. Both aspects are collectively referred to as piezoelectricity or the piezoelectric effect. But how does this lead to a clock?

Cady discovered and quickly patented [7] the fact that he could build very stable resonators using small pieces of quartz driven by a voltage source whose polarity switches with variable frequency. The basic idea is the same as that of a swing being pushed: at a specific frequency of excitation, both quartz and swing move with maximal amplitude. In other words, the quartz exhibits the kind of resonance behavior that was described in the previous chapter for the harmonic oscillator. Compared to mechanical oscillators like tuning forks, suitably cut quartz crystals have an extremely narrow frequency range to which they respond with large amplitude—they exhibit a sharp resonance. It is as if the swing only gets moving at one very specific rate of pushing; at all other frequencies, the swing remains essentially at rest. In addition to being well defined and highly reproducible, the resonance frequency of quartz crystals is also insensitive to temperature fluctuations and can be precisely designed. Quartz oscillators are still used in a large variety of electronic equipment, including clocks, radios and cell phones.

It might seem strange that a solid cylinder would oscillate and have a resonance. Buildings and bridges vibrate because they consist of discrete parts that are somehow connected to each other. But a solid cylinder is a single thing. Of course, this is not really true, since no substance is completely rigid. Any material object is a collection of small masses (in the case of quartz, for example, they are associated mostly with silicon and oxygen atoms) connected to each other by elastic forces, the chemical bonds that hold atoms and molecules in their equilibrium positions. While certainly only a crude image, you can profitably picture the interior structure of stuff as a complex network of masses attached to tiny, stiff springs—a network of simple harmonic oscillators. So, of course a quartz crystal can oscillate. And because quartz and other piezoelectric substances are made up of molecular units that respond to applied electric forces by minute length changes, alternating voltages set up vibrations in the quartz crystal. Naturally, a chunk of matter is not as straightforward as a single simple harmonic oscillator. The main consequence of the coupling of many oscillators is the fact that a given piece of any substance, including quartz, has many frequencies at which it can move: it has a so-called resonance spectrum

---

[2] Many materials exhibit piezoelectric behavior, including certain salt crystals and some organic materials. Even a few biological substances, silk and collagen among them, are piezoelectrically active.

whose exact shape is determined by, among other parameters, the temperature, size and geometric shape, and how the crystal is cut relative to its symmetry axes. For use in many watches, quartz crystals are shaped into the form of a tuning fork (see figure 3.3) with a value of 32 768 Hz for their dominant resonance frequency. This value is 15 times divisible by two, in other words $32\,768 = 2^{15}$. Therefore, a 15 bit digital counter reaches its maximum—all bits are 1—at the end of every second. With the next quartz cycle, the counter overflows, as the jargon goes, and is reset to all zeros. In turn, this condition is used to increment the second count of the quartz clock by one. Such digital counters effectively act as the gearing mechanism of the generic clock discussed in chapter 2. In principle, a 14- or a 16-digit number for the quartz resonance frequency would have worked as well. However, 16 kHz is somewhat annoyingly audible and 65 kHz is just a bit more expensive than is necessary for most commercial watches. Which brings us to the next section.

3.2 ... to quartz clocks

Possibly, the following sentence does not make much sense to you. 'A governor is a part of a machine by means of which the velocity of the machine is kept nearly uniform, notwithstanding variations in the driving-power or the resistance.' Exchange *cruise control* for *governor* and *car* for *machine* and clarity should

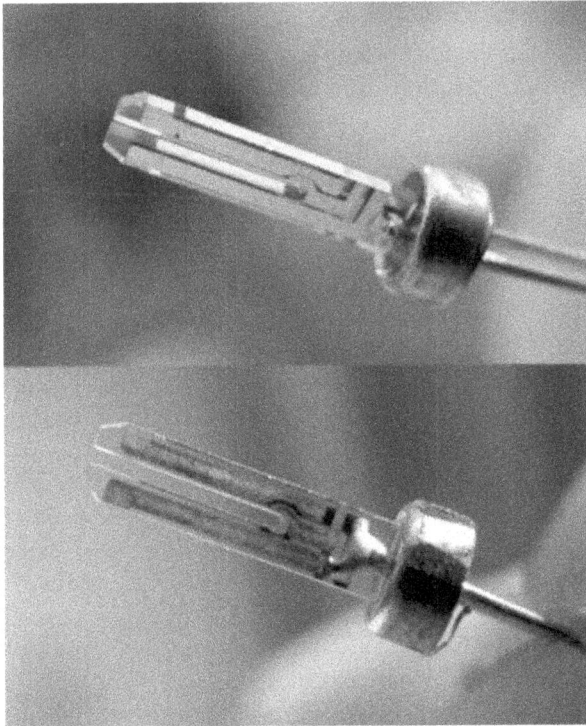

Figure 3.3. Quartz crystal resonator, used as the time-keeping component in quartz watches and clocks, with the case removed.

improve. Since a car *is* a machine, that word substitution seems unproblematic. But what is a governor in this context? The sentence above was written in 1868 and deals with a crucial part of steam engines [8] that prevents them from running at excessive speeds and possible self-destruction—a machine fate not unheard of in the early days of the industrial revolution. Separated by a century, the two pieces of engineering—governor and cruise control—have in common that they are feedback systems. In general, a feedback loop regulates certain processes in dynamical systems. A suitable output is measured and the corresponding quantity is fed back as input into that part of the system that generates the signal in the first place. If dampening of the behavior in question results, the feedback is called *negative*; if the detected signal grows in strength, the feedback is deemed to be *positive*. In clocks, feedback is generally used to adjust the time-keeping mechanism with the goal of the highest possible uniformity of rate of oscillation.

In section 2.8, the issue of feedback has been looked at in the context of resonance. For quartz clocks, resonance and feedback is a marriage of electrical and mechanical components and properties on the basis of the piezoelectric effect. At a given electrical stimulus, aka alternating voltage, the quartz tuning fork vibrates with increasing amplitude when the frequency of the voltage approaches the resonance of the quartz oscillator. On the other hand, a larger amplitude of the mechanical oscillation induces a larger voltage in a separate pick-up coil that generates the necessary feedback signal. An appropriate electrical system—based on either discrete analogue components or integrated circuitry—turns this give and take into a cruise control for the quartz watch. Considering how simple and inexpensive quartz controlled clocks are, they do a marvelous job.

3.3 The first atomic clock was not one

You never know what a substance might be good for: silicon for solar panels, lithium for batteries, or carbon for nanotubes. But ammonia for a clock? Yes, the NH_3 molecule is at the heart of the first atomic clock ever built. The birth certificate of the ammonia clock [9] indicates that it started to run on 6 January 1949 at the National Bureau of Standards (NBS), the forerunner of the National Institute of Standards and Technology (NIST). At its core, a quartz clock was ticking away the microseconds while vibrations of ammonia molecules served to correct small imperfections in the rate of quartz oscillation.

Already in 1879, expanding on a communication he had earlier with J C Maxwell (the same physicist we met above in the context of steam engines and electromagnetic equations) on the subject [10], William Thomson—later Lord Kelvin—remarked that atoms would be an ideal natural time standard, since they are 'absolutely alike in every physical property' and 'probably remain the same, so long as the particle itself exists.' How atoms and molecules, including ammonia, can serve as timekeepers will be the next part of our story.

It is not just cat owners who know that ammonia is a smelly, colorless gas. As noted above, it has the simple chemical makeup NH_3, with the three hydrogen atoms (the white spheres in figure 3.4) at the base and the single (blue) nitrogen

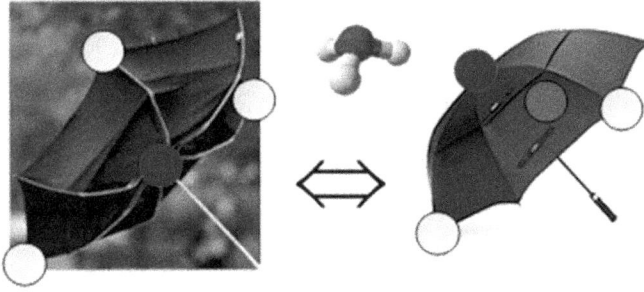

Figure 3.4. Umbrella inversion vibration mode of the ammonia molecule.

atom at the vertex of a trigonal pyramid. That is all fine, but how can a gas become a clock? For starters, the atomic clock in question is made of durable stuff like stainless steel; ammonia molecules 'only' serve as a timekeeper, as the central oscillator. The structure of the ammonia molecule is not rigid and the atoms can vibrate and twist relative to each other and one of the various modes of vibration is exploited for the ammonia clock. If you have ever opened an umbrella in strong wind, you may have an intuitive notion of what the so-called umbrella mode of vibration refers to: the flip-flopping of the hydrogen-containing base from one side of the nitrogen to the other. Or, if you prefer to think of the hydrogen atoms as sitting still, imagine the nitrogen atom performing repetitive inversions from above to below the base of the pyramid. A mechanical model of the NH_3 would have the chemical bonds exchanged for elastic springs strung between tiny masses equivalent to those of the present atoms. The stiffness of the spring is such that the umbrella inversion mode has a vibrational frequency of about $2.38\,701 \cdot 10^{10}$ Hz, almost 24 billion cycles per second—an enormously rapid oscillation that would be impossible to achieve with any macroscopic object. All ammonia molecules are perfect clones[3] of a single blueprint and hardly any impediment slows down their oscillations. As we saw in chapter 2, these two facts together augur well for the quality of a clock that can use ammonia molecules as its timekeeper. But, as we have also seen, a clock needs more than a timekeeper. Specifically, it requires a way to read out the oscillations of the timekeeper. And *that* is neither a trifle nor an easy thing to accomplish here. The essential ingredient turns out to be electromagnetic waves.

3.4 Dancing electrons are making waves—the old-fashioned way

Wave your index finger quickly back and forth through the air and you make a sound. If you do not hear anything, it might be that you are not wiggling fast enough. The low-frequency threshold for human hearing is about 20 Hz, a low-pitch hum. Still nothing? Try a steady rate of 262 wiggles per second, which is close to the middle C. You will have better luck with a tuning fork. Nevertheless, if you could

---

[3] There is one caveat. Chemically identical substances can differ in their mass because they contain different numbers of neutrons in their nucleus. They are called isotopes and play an important role in radioactive decay, which we will discuss in more detail in chapter 5. Different isotopes have different vibrational frequencies.

wave your finger at the same rate as the tuning fork, your finger would generate the same pitched sound wave. Oscillations of a solid object create local compression and rarefaction of the adjacent air volume and these pressure modulations travel, at the speed of sound, away from the source of disturbance. While the amplitude of the sound waves diminishes with distance from the source—the sound becomes fainter—the pitch, i.e. the period of oscillation, stays constant. This is exactly analogous to water surface waves: plunge your finger in and out of the water and ripple patterns will move away in concentric circles. Two corks at two different distances will bob up and down at the *same* rate, but with an amplitude that becomes smaller the further the cork is away from the wave source. By the way, there is one subtle but important difference between sound and surface waves. In the case of the latter, the disturbance—the increase and decrease of local water level relative to the quiescent state—is perpendicular to the direction of propagation. Such waves are called transverse waves. For sound waves, deviations from average density or pressure occur parallel to the direction in which the wave travels.

Because sound waves are pressure or density waves, they need a material medium to be able to propagate. The medium does not have to be air or a gas, it can be a fluid like water or even a solid like the crust of the Earth. But it must be something. Klingon battle cruisers do *not* make a sound in the vacuum of outer space— Hollywood movies notwithstanding. When you search the internet using the cryptic combination 'alarm clock bell jar', the YouTube videos that pop up bear witness to the silencing effect of vacuum. If that is so, what happens to the miniature alarm clock that is the ammonia molecule when it vibrates inside a vacuum environment? We have seen that the frequency associated with this vibration is many orders above anything we or any other animal can hear, but we can still become aware of this vibration because it emits radio waves, the same kind that a suitably modified radio receiver could pick up. What exactly are radio waves? How are they created? First, radio waves are only a special type of a more generic class of waves called electromagnetic (EM) waves, which are emitted whenever an electric charge accelerates. Thus, there is yet another type of wave you can produce by wagging your finger. After dragging your feet over a carpet, it may happen that you are literally charged up and that you feel a small jolt when you touch a door knob. Static electricity has gotten the better of you. In that case, shaking your hand will not get rid of the electric charge. Instead, your moving fingers become a source of EM radiation.

You cannot have one without the other: no accelerated charges, no EM radiation, and when EM radiation is present, electric charges must have accelerated somewhere. The intimate relation also works in reverse in the sense that electric charges exposed to EM radiation start to accelerate, i.e. they experience a force. Newton's second law in action! If an electric charge oscillates with a frequency f, then the emitted waves will oscillate with that same frequency. Pick a random radio station, say Wesleyan University's own student run station in Middletown, Connecticut (WESU). WESU generates radio waves by sending electric AC currents at 88.1 MHz through a broadcasting antenna, which is therefore the carrier frequency of the station's radio waves. Tune in if you are in the neighborhood. (By the way, it is all the way left. Of the dial.) At a microscopic level, electrons are often responsible for doing the moving as they slosh back and forth against a background of heavy positively charged ion cores that remain essentially stationary. In the present case, though, the motion of the whole atoms takes the electrons for a ride. Therefore, if the internal charges were distributed such that each atom in the NH_3 were electrically neutral, no electric wave could or would be produced. However, the central nitrogen atom in the real ammonia molecule attracts electrons a small distance away from the hydrogen atoms, with the effect that the latter are slightly positively charged and the nitrogen atom is slightly negatively charged. This gives the ammonia molecule what is called a permanent electric dipole moment, aligned with its symmetry axis and pointing from apex to base of the pyramid. As is often the case, the details are more complicated but for our purposes this image is sufficient. When the molecule vibrates in the inversion mode, its dipole moment periodically changes direction—the associated electric charges perform harmonic oscillation, thus they accelerate[4] and emit EM waves. In general, EM waves are the direct result of electrons dancing about, whether they do it in the form of an electric current injected into a macroscopic piece of wire, the jiggling of vibrating dipolar molecules, or the commotion of excited atoms. In contrast to sound waves, EM radiation can travel through empty space. Ammonia molecules have been found that, unlike Klingon ships, *actually* cruise in outer space: the characteristic ammonia inversion signal at 23.87 GHz has been detected emerging clearly from molecular clouds in our galaxy [11].

Many types of EM waves exist, with frequencies spread out over many orders of magnitude. Figure 3.5 illustrates the EM spectrum for more than 13 orders of magnitude in frequency f and wavelength, usually denoted by the Greek letter λ. The relation between frequency and wavelength is most easily seen by considering surface waves, but the result is valid for any type of wave. Suppose a wave train travels from east to west on a body of water, with two adjacent crests separated by the wavelength λ of the wave. Several pieces of cork in the water act as motion detectors, bobbing up and down with each passing wave crest and trough. All of the corks move up and down with the frequency of the wave, which is also the frequency of the source oscillation that created the wave (say your finger dipping regularly in and out of the water). If we arrange the distance between two of these corks to be equal to the wavelength λ of the wave, the motion of these two corks will be synchronized—they are simultaneously all the way up, down at the bottom of a crest, halfway between, etc. Now pay attention to the motion of a single crest as it moves from the first to the second cork. What is the speed of this crest? We already know that it travels a distance λ between the corks. Thus, the speed is known when we know the time it takes to move this distance. When the crest starts at cork 1, cork 2 is also at its highest point of motion. And when the crest in question arrives there, cork 2 is again at its highest point. In other words, it has completed one cycle, one period of oscillation of the wave. By definition, period T and frequency f are the inverse of each other, so that the wave speed v is found to be

---

[4] If you need to refresh your memory, check out the discussion of harmonic oscillators in chapter 2.

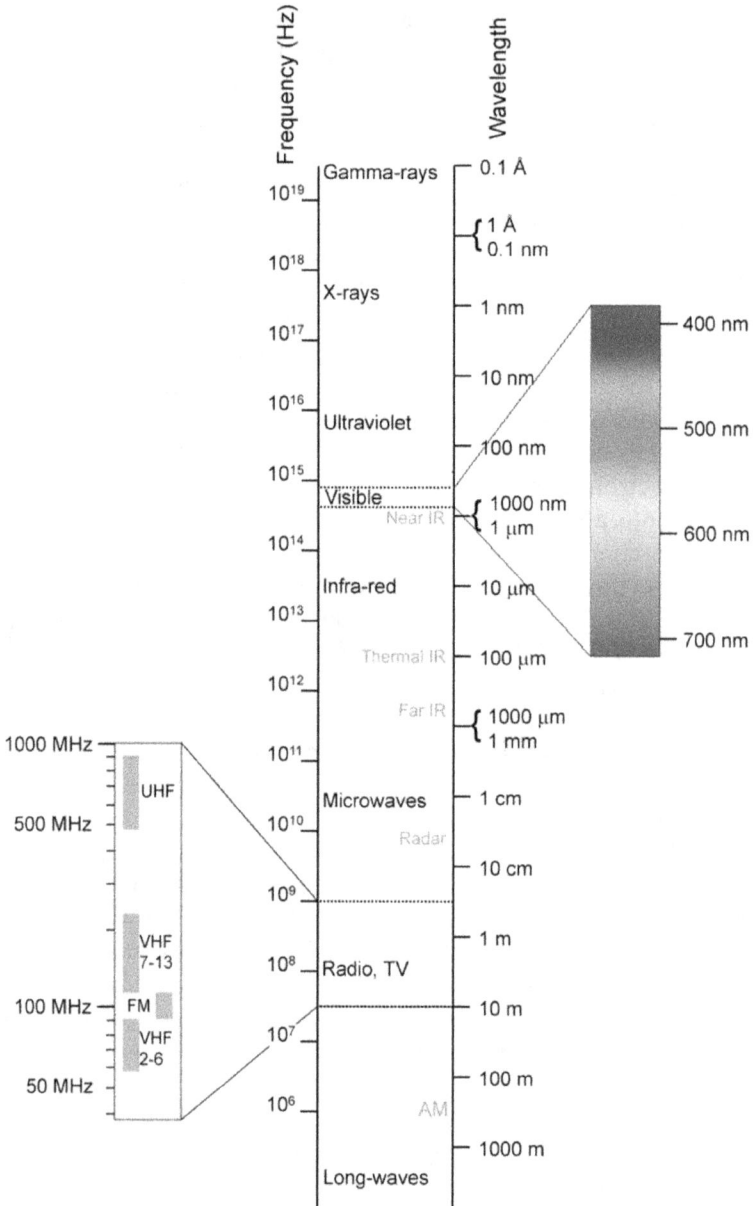

Figure 3.5. Spectrum of electromagnetic (EM) waves.

$$\{\text{wave speed}\} = v = \frac{\{\text{distance traveled}\}}{\{\text{time needed}\}} = \frac{\lambda}{T} = \lambda f.$$

In vacuum, the speed of EM waves is customarily written as c and has been measured to be about $300\,000$ km s^{-1}. With this enormous value, a laser pulse races to the Moon in a tad more than one second. The above relation and the value of c

are all that is behind the conversion between frequency f and wavelength λ shown in figure 3.5. As you can see, the range around the ammonia inversion frequency coincides with microwaves, the short wavelength cousins of radio waves. To the long-wavelength or low-frequency side, the spectrum extends beyond the radio waves. Radiation with frequencies from 10 to 100 Hz or so are unimaginatively called extremely low-frequency (ELF) waves, with corresponding wavelengths reaching tens of thousands of kilometers. ELF waves are generated by lightning and other natural electric disturbances in the atmosphere, but also by power grids running at 60 or 50 Hz. In principle, even lower frequency EM waves might exist. EM waves in the visible part of the EM spectrum play a role in so-called optical clocks. Because of the higher operation frequency, they have the potential to replace the current cesium clocks as time standards. At the moment of writing (2017), prototypes exist that have shown great promise to replace the question mark in the top of figure 3.1 by a new item N. Beyond the visible and UV range, x-rays beckon as the new frontier. Given the vast range of high frequencies available (see figure 3.5), it is unlikely that the quest for more accurate clocks will end any time soon.

For more than 100 years, scientists and engineers have added chapter after chapter to an ongoing and tremendously successful story of generating, detecting and controlling EM waves at a steadily increasing range of frequencies. Telecommunications and data transfer in particular have benefited tremendously from the opening of new spectral windows in the higher frequency regions. These achievements are based on the theoretical understanding outlined above in broad strokes. We will return to the issue of generating EM waves in the section on the quantum revolution below. The discussion presented so far in this chapter recounts the 'classical' or pre-quantum story line, which is flawed at the fundamental level but, under the prevailing conditions, gives results that are correct to a high degree of approximation. In the quantum section, we will also obtain a glimpse of how the two points of view can be reconciled.

3.5 Inner workings and limits of atomic clocks

The 21 January 1945 front page of the *New York Times* announced that Roosevelt had been sworn in for his fourth term as president of the United States. Further inside, a short article on page 34 was concerned with science news. Under the flashy title '"Cosmic pendulum" for clock planned' [12], the piece reported designs 'for the most accurate clock in the Universe' that had been revealed at a meeting of the American Physical Society by Isidor Isaac Rabi (or I I Rabi, as he preferred). Apparently, the *New York Times* assumed that no other intelligent, or rather clock-making, life form existed anywhere in the Universe. To be sure, the proposed timepiece *was* revolutionary and marked the beginning of a stunning development in clock-making that gave rise to the type of clocks labeled K through M in figure 3.1. The *New York Times* article went on to describe the underlying concept as being based on 'minute radio transmitting stations in the nuclei of atoms' as well as 'exceedingly delicate receiving sets for tuning in on these cosmic broadcasts.' In 1944, I I Rabi had received the Nobel Prize in physics [13] for the discovery of the former and the development of the latter, and both items play crucial roles in atomic (and molecular) clocks. Soon after the invention of the ammonia clock, which never could be made to run accurately enough to be used as a time standard, a new breed of now truly atomic clocks was introduced, based on a radio signal of 9 192 631 770 cycles per second associated with cesium. Ammonia and cesium clocks are conceptually very similar, even if they differ in the details of how their 'minute radio stations' operate. Below, we will return to this latter aspect, since the differences explain why one clock, the ammonia clock, never ran accurately enough to serve as a time standard while the other, the cesium clock, did. For now, we concentrate on the unifying essentials that make atomic clocks tick.

In the context of atomic clocks, it is vaguely misleading to speak of 'radio *transmitting* stations in the nuclei of atoms'. That sounds as if we would listen to the programs emitted by the atomic (or molecular) radio stations. In reality, the atoms (or molecules) act as an eager audience, all tuned in to the same show that is being sent to them. More accurately still, it is an eager station manager who tweaks the broadcasting frequency of her station until all of the radio receivers, whose circuits are set to the same frequency, receive the signal. Thus, in atomic clocks, the role of who does the tuning is reversed compared to the normal radio broadcasting case. If WESU were to adopt this idiosyncratic, not to mention expensive, approach, how would the station manager know if the available electric power, about 6000 watts, is converted to EM waves with just the right frequency to ensure optimum signal reception at the fixed frequency receiver end? Here is an idea[5] that, for a variety of reasons, will definitely not work in practice. Measure the power absorbed by all receivers tuned to 88.1 MHz. Then change the frequency ever so slightly and see if the absorbed power increases or decrease. If it decreases, the broadcasting frequency was already optimized, so you must return to the previous frequency value. If the power loss due to absorption increases, keep going until it decreases. You just passed the optimum. Continue to monitor power loss at all times. In this fashion, the broadcasting frequency can be first tuned into and then kept at resonance with the receivers. Believe it or not, but that is exactly the way atomic clocks keep their frequency—in this circumstance, the method does work. Like a charm.

The flow diagram in figure 3.6 shows schematically how such feedback and control works for a generic atomic clock. Everything used in a quartz clock—driving mechanism, frequency counting and so forth—can be recycled. The only and essential addition is the use of atomic or molecular frequencies as the timekeeper. Any small drift, erratic or systematic, of the quartz oscillator is corrected by actively locking it to the much more stable atomic frequency chosen. Of course, the flow diagram in figure 3.6 does not reveal the significant engineering challenges associated with each separate clock component that had to be overcome before atomic clocks became the high-precision tools they are today. Nevertheless, the figure provides an adequate representation of the basic inner workings of atomic

---

[5] Apparently I was not the first—I rarely, if ever, am—to think of this: see [14].

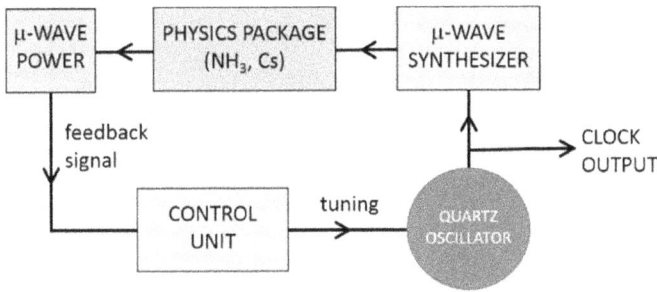

Figure 3.6. Atomic clock flow diagram.

clocks. A local oscillator in the form of a quartz crystal determines the frequency of the microwaves that are sent through a vacuum tube containing the ammonia molecules or cesium atoms. Therefore, the oscillator quartz must be cut and sized so that its resonance frequency coincides, within an integer factor, with the resonance frequency of the specific microscopic timekeepers used in the clock. When the frequency of the source that generates the EM waves, the microwave synthesizer, is adjusted to the maximum of the absorption curve, maximal power loss occurs and is sensed by a detector. Deviation from this optimum frequency—determined by the substance used in the atomic clock, cesium these days—can only occur in an extremely narrow range, outside of which the absorbed power drops significantly. Information on the absorption levels is fed back into sophisticated electronic circuits that evaluate the significance of the measured signal and initiate appropriate action to keep the quartz oscillator frequency locked to the atomic or molecular resonance. There are two key ingredients for the remarkable performance of atomic clocks. First, these marvels of technology require state-of-the art engineering. Thousands of scholarly papers, patents, textbooks, conference proceedings and other forms of technical communication document the impressive evolution that took place over the past 70+ years. Equally important though is a simple physical phenomenon— the sharp resonance curve of the system (see section 2.8). Even if the clock were constructed using the best engineering imaginable, an intrinsic physical property of the microscopic timekeeper sets the ultimate performance limit. We will discuss related aspects of this issue in more detail in the sections below on the quantum nature of atomic clocks and clock comparison. Here, figure 3.7 will suffice to clarify the point. The diagram shows measurements performed for the Brazilian[6] time and frequency atomic standards program characterizing a new type of cesium clock [15]. Ignoring all details, the absorption of cesium atoms—expressed as a *transition probability* in arbitrary units—exhibits a narrow peak at a specific micro-wave frequency, which is quantified as detuning away from the optimum for stationary atoms. Thus, the zero point of the x-axis corresponds to 9 192 631 770 Hz. A fit of the known theoretical function (dotted smooth curve) to the measured

---

[6] Timekeeping with atomic clocks is a truly global enterprise.

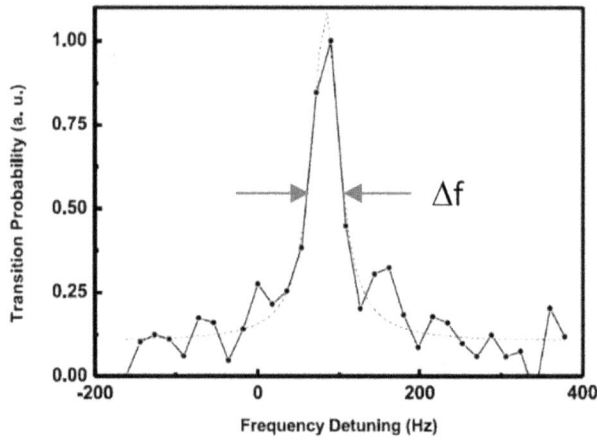

Figure 3.7. Cesium resonance curve. Reproduced from [15] with permission.

data (symbols connected by line segments) reveals that a detuning to either side of the maximum by about 20 Hz reduces the absorbed power by a factor of two. Therefore, how well this particular set-up would function as a clock also depends on the ability to detect small changes in absorbed power. For example, if the feedback loop of the clock can detect a change in absorbed power that corresponds to a frequency drift of ± 10 Hz, the clock frequency can stray by no more than that from its optimum.

Since a sharp resonance curve is so important, it is useful to ask which factors determine the sharpness of a given system. Let us first remind[7] ourselves of the relation between the Q factor, which is the parameter that summarizes the resonance sharpness, and clock accuracy, as depicted for example in figure 3.1. In the simplest approach, we may write $Q = f/\Delta f = T/\Delta T$, where Δf is the width of the resonance curve, f is the oscillator frequency at resonance, i.e. at the maximum of the curve, and ΔT is the typical error in the corresponding period T of the oscillation (remember that $T = 1/f$). This latter uncertainty propagates forward into an uncertainty Δt of measuring a duration t. In figure 3.1, for example, the time duration was chosen to be one day. If we express t as a multiple of the base period T, i.e. $t = NT$, then also $\Delta t = N\Delta T$, so that furthermore $Q = t/\Delta t$ and therefore $\Delta t = t/Q = t \cdot \Delta f/f$. This relation implies that accuracy of time measurement can be achieved in two ways. For a given type of clock, i.e. for a given value of the resonance frequency f, decreasing the frequency spread, i.e. increasing the sharpness of the resonance, will improve the time measurement uncertainty Δt. However, for a given frequency uncertainty Δf, it is also possible to enhance clock performance by increasing the resonance frequency. This strategy requires a change of either the atom or the type of transition of a given atom that is used in the clock. Since the invention of the cesium atomic clock, the first method has been the *de facto* approach. However, intensive efforts to measure directly frequencies in the visible light range of the EM spectrum,

---

[7] See chapter 2.

where transition frequencies are three orders of magnitude higher than the current cesium standard (see figure 3.1), have borne fruit. It is likely that in the near future [16] the definition of the second will be connected to optical transitions of either the strontium or the ytterbium atom and that corresponding atomic clocks will come online.

Whether cesium, strontium, or some other chemical element is used, in the foreseeable future the most precise clocks will employ atoms as their timekeeper. More specifically, this type of clock relies on the motion of the electrons that constitute the outer skin of atoms. Very speculative ideas explore the possibility of tapping into the oscillations of the nucleus, the pit of the atom, which might open up the realm of *nuclear clocks* [17]. Conceivably, they could tick at rates that are several orders of magnitude higher than even the optical clocks currently on the horizon.

3.6 How time fared in the quantum revolution

'Traditional' or up-and-coming, atomic or optical or even nuclear—all these clocks transfer the act of time-keeping from a level of direct observation to scales of extreme abstraction. While it is true that the definition of the second afforded by these high-precision clocks uses natural objects and processes and thus still adheres to the spirit of the unit system born in the French Revolution, it is far removed from everyday experience. Our 'sense' for the solar and the sidereal day or for the swing of a pendulum is firmly based on direct observation and is thus amenable—for better or worse—to intuition. This is of course not the case for anything that plays out on the stage of molecules, atoms, or nuclei. So, we ought not be surprised that the rules of classical physics, which we deduced from studying macroscopic phenomena, are not applicable to the realm of the microscopic. Why should they be? Collectively, these new rules are known as quantum physics and were established, for the most part, during the first three decades of the 20th century. It goes without saying—yet I do it anyway—that the history, scope and complexity of quantum theory far exceed the limitations of this slim treatise. We can be content with snippets of its chronicle and a small sampling of quantum weirdness as it pertains to clocks and time. Somewhat arbitrarily, but with an eye towards aspects that are important for a better understanding of atomic clocks and fundamental limits, I have chosen to concentrate on the following aspects: (1) the existence and permanence of lowest energy states, the so-called ground states of atoms, molecules and nuclei; (2) the unavoidable presence of fuzziness, characterized by equations known as Heisenberg uncertainty relations; (3) the role of randomness in the measurement process. Before we start, here is a warning from Richard Feynman, someone who should know: 'I think it is safe to say that no one understands quantum mechanics. ... Do not keep saying to yourself, if you can possibly avoid it, "But how can it be like that?" because you will get "down the drain" into a blind alley from which nobody has yet escaped. Nobody knows how it can be like that' [18]. With that in mind, let us avoid the pull of the vortex, lurking at the center; let us stay at the somewhat safer periphery.

3.7 Inside the hydrogen atom

With every sip of refreshing water, you absorb hydrogen atoms into your body that are far from being fresh—they were made more than 14 billion years ago[8]. In the last chapter of this book you can find out how we know. Astonishingly, despite their remarkable age, the atoms have not changed a bit—as far as we know, they do not have an expiration date. None of this would pose a conceptual problem if Anaxagoras, Leucippus and Democritus had been correct. Living, thinking and teaching in ancient Greece, these three philosophers are usually credited with an early if not *the* earliest proposal that all matter in this world is made up of a set of smallest, indestructible atoms (Greek: *atomos*—undivided). It is not that they had any empirical evidence—that came much later—but the other notion, namely that there is *no* limit to the divisibility of matter, is equally eccentric. So why not a Lego set of tiny spheres, cubes, pyramids ... what have you. Of course, when you conjure up the image of a small, but nevertheless finite-sized object of any shape, you are tempted to ask what the inside looks like. For Democritus, the inside of atoms was hard, motionless, without feature and eternal. He and his teacher Leucippus had figured out how the world hangs together. Still, the atomic idea did not catch on until more than 2000 years later. When, for example, John Dalton noted simple integer relations in the consumption of reagents in chemical reactions. Or when Dmitri Mendeleev came up with a scheme to organize the chemical elements. Or when Johann Josef Loschmidt estimated the size of air molecules (he was off by only a factor of two—quite impressive, considering what was available). By the time J J Thomson discovered the electron, in 1897, and suggested that it was a light building block of the atom, the existence of the latter was widely, though not universally, accepted in the scientific community. Only a dozen years later, Ernest Rutherford explained an experiment in terms of a model of the atom that is valid to this day. He had proposed to bombard gold foils with fast particles emanating from radioactive substances (see chapter 5) to learn more about the inner structure of atoms. In 1907, Hans Geiger and Ernest Marsden had carried out this experiment successfully. There was no doubt that electrons were indeed part of the atom, but they had to be on the outside. At the center, a small, compact pit—the nucleus—had to reside something whose mass made up almost the entire mass of the atom. We now know more precisely that the smallest nuclear constituents, the positively charged proton and the electrically neutral neutron, are a bit more than 1800 times heavier than the electron[9]. So, there you have it. Two thousand years after Democritus, we know what is inside the sphere that is the hydrogen atom: an electron somehow moving around a proton. And therein lies the problem.

According to the rules of physics known in the first decade of the 20th century, hydrogen atoms cannot be stable. Electron and proton strongly attract each other by virtue of their electric charges. The only way to balance this attraction is the same

---

[8] To be a perfect stickler: the age refers to the proton. The electron is even a bit older.

[9] Using the currently accepted mass values, the ratios between proton and neutron mass and that of the electron are $1836.15\,267\,389\pm17$ and $1838.68\,366\,158\pm90$, respectively. See e.g. [19].

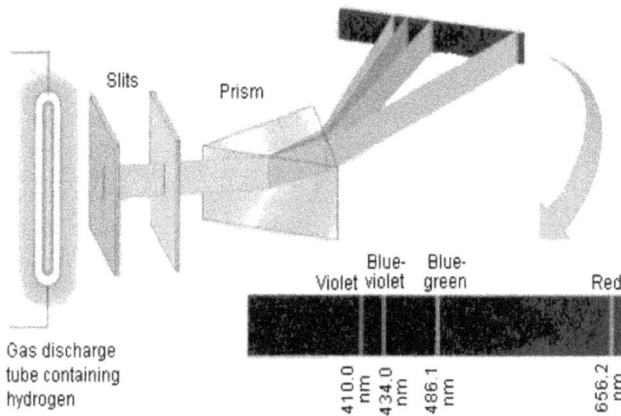

Figure 3.8. Schematic set-up for measuring the emission spectrum of hydrogen. Reproduced from [20] with permission of John Wiley Sons.

way planets ward off the gravitational pull of the Sun. They must circle the center. But if an electrically charged particle does that, it emits EM waves (see above) that carry energy away from the electron. This slows the electron down until it crashes into the proton. A back-of-the-envelope calculation shows that the time span for the demise of an electron that starts at the radius of the hydrogen atom, about 0.05 nm, is measured in nanoseconds. Hydrogen atoms should vanish in a flash. How come they are still around? Enter Niels Bohr.

Niels Henrik David Bohr was 28-years old when he made a simple observation: what is not allowed must be forbidden. Bohr *postulated* the stability of the hydrogen atom and then followed the path that opened when he *assumed* that some yet unknown property prevents undisturbed atoms from emitting EM waves. The qualification of isolation is necessary because hydrogen atoms do emit light when, for example, bombarded with an electric current (think fluorescent light tubes). But once left alone, the light show subsides and the atoms turn dark. Dispersion of the light emitted upon stimulation with a prism reveals that it contains only a few very narrow bands of color (see figure 3.8), whose wavelengths were known to five or even six digits—a respectable level of precision[10]. Bohr's ad hoc model of the atom, published in 1913, also reproduced these wavelength values with eerie fidelity. He had to be on to something. So, what exactly were Bohr's postulates? Another detour is in order.

3.8 Quanta make their entrance

When Bohr worked on the hydrogen atom, he was one of the early protagonists of the quantum revolution that was about to shake the foundation of physics. In 1900, this revolution had started quietly with a seemingly mundane publication by a

---

[10] We have come a long way, though. At present, the frequency associated with a hydrogen emission line in the UV part of the spectrum (the $2S$-$2P_{3/2}$ transition) has been measured to almost 15 digits of accuracy [21].

physicist named Max Planck. He had worked out a theory that provides a microscopic explanation for the well-known observation that objects look alike when they become hot. Coal, glass and iron—transparent, black and metallic at room temperature—all turn first dull red, then bright orange/yellow, and then bluish white when they turn sufficiently hot. Planck's mathematical formula neatly replicated the observed color spectrum of 'black bodies', as the idealized cousin of coal, glass, iron and any other substance is called in physics lingo. So far, so good. What annoyed Planck was a perceived flaw that kept showing up in the physical models he employed. He treated the emission and absorption of light by modeling the atoms of the material as tiny oscillators, very much in the spirit of the above quoted *New York Times* article on atomic clocks. The defect of his theory, as Planck and most everyone else saw it, was the detail that these oscillators had to absorb and emit electromagnetic radiation in chunks. Only then could he obtain the radiation formula he needed to describe the observations correctly. Specifically, the energy quantum had to be proportional to the frequency f of the oscillator. Planck labeled the proportionality constant, now known as the Planck constant, with the letter h, a label that stuck and that is used with universal consistency. In all of physics before 1900, energy was a continuous quantity. In the new theory of thermal radiators, the currency of energy emitted and absorbed by atoms has a smallest coin with a face value equal to hf—energy is quantized. Planck firmly believed that this was a mathematical artifact that, with sufficient patience and skill, could be made to vanish. It turned out that he was wrong. The physics community either did not take notice of this irritation or shared Planck's view on the matter—with one notable exception. In 1905, Albert Einstein not only accepted the idea that atoms exchange electromagnetic energy in quanta of magnitude hf, he spun Planck's yarn even further and asserted that light, electromagnetic radiation, itself existed in quantized, smallest denominations of hf. Far from being just a speculation, the light quantization idea allowed Einstein to explain in detail all the features of the photoelectric effect, the summary name for the phenomenon of the ejection of electrons from illuminated metal surfaces. Much later, the name *photon* was coined for this smallest energy increment of light that Einstein had introduced. It is amusing that he received the Nobel Prize in physics for this highly successful application of the quantization idea—amusing because, later in life, he turned from an early supporter to a persistent detractor of quantum physics.

Now we are ready to return to Bohr's contribution, which was recognized by the 1922 Nobel Prize in physics 'for his services in the investigation of the structure of atoms and of the radiation emanating from them' [22]. Even if quantum physics has evolved since Bohr's first papers into a much more sophisticated theory, his hypothesis remains a useful launching point to discuss light–matter interaction at the atomic level. The essence of Bohr's three postulates can be expressed, loosely, in the following way. (1) Electrons inside atoms exist in discrete, quantized 'orbits' or 'states' with different sizes and shapes and thus with different energies. Among these quantum states, one has the lowest energy possible for a given type of atom. It is called the ground state; all others are referred to as excited states. Barring any external energy input, the ground state cannot lose energy and is thus perfectly stable. (2) Atoms can only absorb or emit energy in the form of photons if the energy of the latter is equal to the absolute difference of the energies of the two atom states before and after the change, i.e. provided that $hf = |E_{before} - E_{after}|$. (3) Bohr's recipe for constructing a better atom contained a third, crucial ingredient that acts as an orbit filter. It sorts out certain values of the angular momentum, a dynamical quantity in mechanics that characterizes motion along curved paths in general and closed orbits in particular[11]. Of all the possible paths that an electron can take around the proton, the only ones that appear in real atoms are those for which the angular momentum is, apart from a factor 2π, an integer multiple of the Planck constant. Because the numerical factor appears frequently, the quantity $\hbar = h/2\pi$ is in common use as a second form of the Planck constant ($\hbar$ is pronounced 'h-bar'). With this prescription, Bohr introduced yet another role of the Planck constant in creating quanta where classical physics had continua. Using the frequently chosen letter L to denote angular momentum, we can express Bohr's third postulate succinctly as $L = n\hbar$. Apart from the above postulates, Bohr used only the laws of classical, macroscopic mechanics to evaluate the radii and energies of possible orbits of the hydrogen atom. With those numbers in hand, postulate (2) immediately yields the frequencies and thus wavelengths of the light hydrogen atoms should emit according to his model. Lo and behold, as already alluded to, the predicted and measured values matched within experimental uncertainties.

The achievement launched a scramble to find an explanation for the success of Bohr's theory and to find a conceptually more consistent basis. However, already at this preliminary stage, several important aspects are emerging that pertain to atomic clocks. Even if electrons inside atoms seem to behave somewhat like a miniature solar system, their usefulness for time-keeping in clocks is utterly different from that of their macroscopic counterparts. Electron orbits around the nucleus do *not* provide by themselves a means to read off time. There is no period, no 'atomic year' marking the completion of one round-trip that could be measured. Orbital periods emphatically do *not* serve as timekeepers in atomic clocks. If left in the ground state, atoms do not change at all and thus are unusable as clocks. As we have seen above, it is the oscillatory nature of EM waves that drives atomic clocks. Now we know that it is the intrinsic, unchanging nature of the atom that determines the frequency of the waves. Because two orbits are involved in the emission or absorption, we talk in general about transitions. We will see next what determines the ultimate sharpness of the corresponding transition frequencies. In chapter 2, I alluded to the possibility of describing electrons inside atoms analogously to mechanical oscillators. In Bohr's model of the atom, this is not clear at all. The next section will give a hint as to why this analogy is quite reasonable after all.

---

[11] The following points may help to illustrate the concept. (1) Straight line motion has zero angular momentum. (2) For an object of mass m circling a center at distance r and speed v, the angular momentum is given by $L = mvr$. (3) In elliptical orbits, radius and speed are variable and the right-hand side of the equation for the circular case must be replaced by a suitable integral, taken over one period of the orbit.

3.9 Precisely specified fuzziness

Bohr's model of the atom is a hybrid of old and new concepts. It leaves Newtonian mechanics in place, constrained only by the unusual angular momentum selection rule and the no-radiation decree. Consequently, the notion of well-defined paths and trajectories remains untouched. If you live in or visit Amarillo, Texas, you can see an artistic embodiment of this idea applied to the two electrons in the helium atom (see figure 3.9). At the point where four columns meet, a sculpture of a helium atom is suspended. Four spheres represent the two protons and two neutrons of the nucleus. They are surrounded by two rings that symbolize the orbits of the two electrons. This is probably the archetypical image of an atom that most people carry in their mind. Just do a basic internet search of the word 'atom' and see what images turn up. We now know that the picture of smeared-out clouds is a more apt representation of electrons inside atoms than well-defined orbits. How this knowledge of something so remote from our senses came about, is a fascinating story. By the way, recent advancement in atomic force microscopy has made it possible to see individual atoms, at least their outline, shape and size. Democritus would have been happy.

Numerous physicists made important contributions to the foundation and subsequent development and expansion of quantum physics. However, if pressed

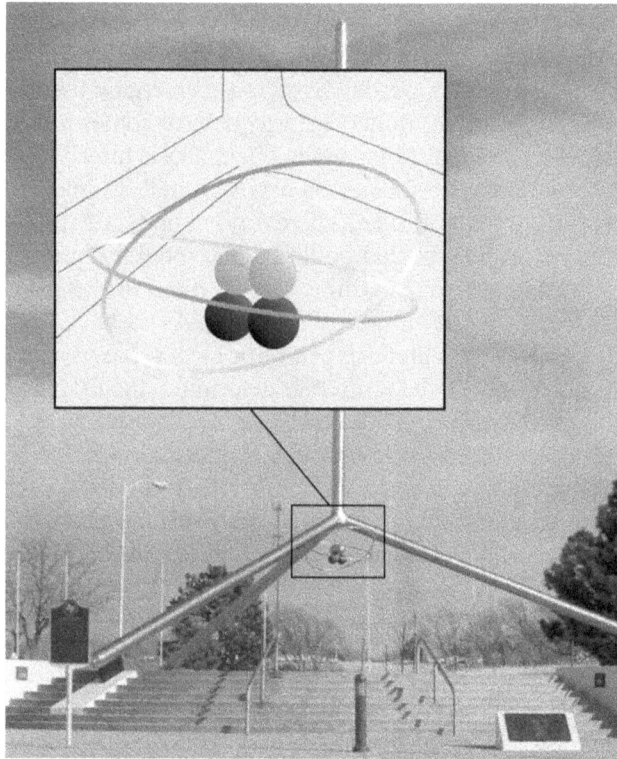

Figure 3.9. Helium Time Column monument in Amarillo, Texas. The two electrons of the helium are represented by rings reminiscent of planetary orbits.

to name only two of the early pioneers, many people would probably come up with Schrödinger and Heisenberg. In a way, they are the Yin and Yang of quantum physics, although Schrödinger's virtual cat has to be the clear favorite for that title. Within a span of less than half a year, the two scholars published frameworks for the microscopic world. Using differential equations, a tool that was very familiar to theoretical physicists, Schrödinger built a bridge from the old continuous world to the new realm of disjointed quanta. Because the central equation in his work is similar to a well-known equation for waves, Schrödinger's approach has been dubbed *wave mechanics*. Quantization of confined systems arises from Schrödinger's wave equation in analogy to the behavior of confined, so-called standing waves[12]. Rejecting any need or possibility for using familiar imagery, Heisenberg introduced a very abstract branch of mathematics, the linear algebra of matrices, to describe the world of atoms. With help from others, this new methodology was soon refined and became known as *matrix mechanics*. Heisenberg's theory represents measurable quantities, so-called observables, by the elements of matrices, tables of numbers that obey specific rules of manipulation, and the corresponding measurements by well-defined mathematical operations involving these matrices. Quantization emerges as an intrinsic property of the mathematical structure. Heisenberg's mantra is that only observables matter and that the theory must be silent about those things that cannot be measured. That includes electron orbits.

Not only did Heisenberg's quantum mechanics have no use for trajectories and orbits, it also gave a deeper reason for this rejection. In classical mechanics, the path of a particle is controlled by the forces acting on the particle and the values of its position and velocity at a single time, which, for the current purpose, we refer to as the present. With that information, as we saw in chapter 2, past and future positions and velocities are uniquely and completely determined (which is of course why Newtonian mechanics is called deterministic). The collective set of all points visited by the particle is its trajectory and can be represented by a line in space, for example the two stylized electron orbits of the Amarillo sculpture in figure 3.9. Note that a simultaneous knowledge of position and velocity is required in the definition of a trajectory. Knowledge of two quantities implies their existence. But, according to Heisenberg, that exact prerequisite is not fulfilled. Instead, for a given particle with mass m, the two quantities of position x and velocity v are only defined to within a spread of Δx and Δv of the corresponding quantities[13]. The two quantities are linked by the relation $\Delta x \cdot m\Delta v \geqslant \hbar$, one of the Heisenberg uncertainty relations, which permits that one of the observables can be specified with arbitrary precision, i.e. with a spread tending to zero, but not both. Why is it that classical physics has no such relation? The answer is simple: the absence is based on a faulty assumption, namely the above-mentioned existence of two infinitely precise variables, here position and

---

[12] Strings of musical instruments are a good example, as are the vibrations of drumheads or any taut membrane. In 1787, Ernst Chladni, a musician and physicist, invented a method to render such vibrations visible by sprinkling fine grains like sand or salt onto the oscillating surfaces (see figure 3.10).

[13] Technically, the two complementary quantities are position x and momentum $p = mv$. If the particle behaves non-relativistically, mass is constant and the two forms are effectively identical.

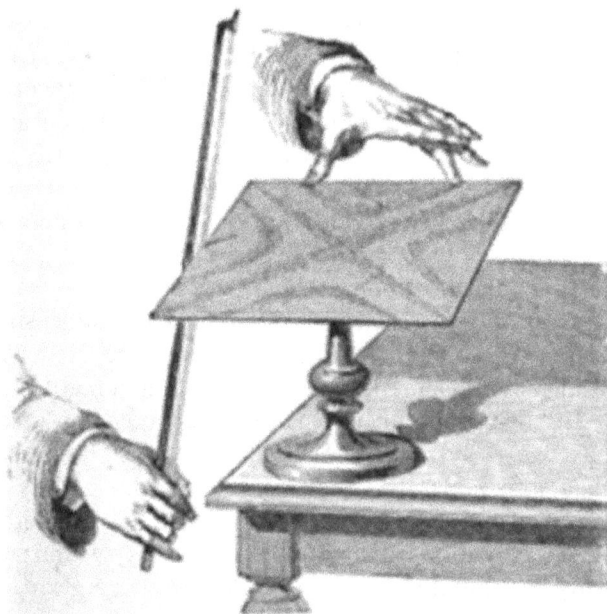

Figure 3.10. Ernst Chladni's method to visualize standing waves on vibrating plates works with fine grains that accumulate along lines of 'silence' of the wave pattern, the so-called nodes, and which get tossed away from the areas of largest motion, the wave crests.

velocity. Implied is the possibility that each of the quantities can be measured without disturbing the other in the slightest. Of course, you can always *assume* that this is possible—at the least, it is a highly intuitive supposition. An example will show how the uncertainty relation plays out in the microscopic world, and will also illustrate why the macroscopic world does not require it.

First, we need to know the value of the Planck constant. Currently, the best value [23] for h is $6.626\,070\,040{\cdot}10^{-34}$ Js which gives, in round numbers, 10^{-34} Js for $\hbar$. We also need the electron mass, which is about 10^{-30} kg. Then, the Bohr model tells us that an electron in the ground state orbit of the hydrogen atom is zipping along at a speed of about $2{\cdot}10^{6}$ m s^{-1} in a circular orbit with a diameter of about 0.1 nm, the typical size of an atom. Suppose a measurement determines the speed of the electron with an uncertainty of 10%, which is sufficiently precise to obtain a reasonable estimate of how the electron is moving. However, in that case, all that can be said about the electron position is that it lies within a range ten times larger than the average orbital diameter. Hardly the precision needed to define a trajectory! In other words, at the atomic scale, Newtonian mechanics does not hold. However, mostly because of the vastly larger masses of even the smallest large objects, such as dust specks or viruses, the uncertainty relations can be completely ignored at the macroscopic level of our daily experience. Apart from position and velocity, other pairs of observables exist that are linked by an equivalent uncertainty relation, for example the angular momentum and angular position of circular motion. Most important for our topic is the combination of energy E and time t, $\Delta E \cdot \Delta t \geqslant \hbar$.

Several interpretations of this relation exist, most notably the association of ΔE with the uncertainty of an energy measurement and of Δt with the time necessary for this measurement. Unfortunately, one aspect of time in formal quantum theory puts a damper on this and similar explanations—in the formal structure of quantum physics, time is *not* an observable like energy, position, velocity, or angular momentum. In other words, time is *not* treated as something that is *measured*, but rather as something that is *given* as a parametric value on which the values of other, measurable quantities depend. It is rather strange, but that is the way it is. Maybe the least controversial way to read the energy–time uncertainty relation is associated with the case of the energy measurement of photons that are emitted from excited atomic quantum states. Here, the energy uncertainty ΔE turns into an uncertainty $\hbar \Delta f$ of the frequency of the photons. If the excited state decays during a characteristic lifetime τ, we can identify the time uncertainty with this time constant and find for the uncertainty relation $\Delta f \geqslant 1/\tau$. In chapter 2, in the context of the discussion of the resonance behavior and the Q value of damped, decaying harmonic oscillators, we encountered essentially the same relation. We may take this as evidence that electrons in confined spaces, for example inside an atom, behave analogously to decaying oscillators. Therefore, the sharpness of atomic resonance curves, such as the one of cesium in figure 3.7, is ultimately limited by the decay time τ. The quicker the decay, i.e. the smaller τ, the broader the resonance curve and the less useful the atom is for a precision clock.

The energy–time uncertainty also highlights the peculiar fact that measuring the energy of an excited state exactly will take an eternity: if the energy should be of ultimate precision, i.e. $\Delta E \to 0$, then it follows that $\Delta t = \hbar/\Delta E \to \infty$. Since the ground state energy is the lowest possible, we may set it as the zero point of the energy scale of that atom. With that choice, the energy of the emitted photon is exactly equal to the excited state energy. Therefore, the excited state energy spread ΔE shows up directly as a spread $\Delta f = \Delta E/h$ of the frequency of the EM wave. This finding implies that atoms with the longest life time and measured for the longest practically possible way should generate the best atomic clocks.

By now, the concept of *measurement* has come up in various contexts. Undeniably, it plays a central role in quantum theory. The next section aims to clarify the role of measurement and the observer as well as the way probability and randomness are involved in all this.

3.10 Does nature play dice?

First, there was no mathematically rigorous and consistent quantum theory. Suddenly, thanks to Schrödinger and Heisenberg, there were two. Because of their mathematical differences, it was not initially clear whether one formalism was in any way superior to the other. In addition, the physical significance of either theory was unclear. At least the resolution of the first question came rather swiftly by a proof of their formal equivalence. In a sense, the two theories told the same story in different dialects of mathematics. As to the interpretation of quantum physics, that was a tougher issue, and some claim it is still not answered satisfactorily—Feynman's above quoted remark comes to mind. After intense debates and with the partic-ipation of many eminent scientists[14], eventually a majority consensus emerged, a view that is expressed in almost all current physics textbooks and that is commonly referred to as the Copenhagen interpretation of quantum mechanics. A word of caution is in order. As Asher Peres puts it: 'There seems to be at least as many different Copenhagen interpretations as people who use that term, probably there are more' [24]. Undaunted—or possibly encouraged—by this verdict, I offer an attempt to summarize the measurement process as described in quantum mechanics.

Depending on the details of their preparation, physical systems (for example single atoms or larger collections of atoms) exist in certain quantum states. Any measurement disrupts the system in a way that cannot be controlled, causing a state change of the system. In general, quantum theory only provides statistical pre-dictions for the possible outcomes of such measurements. For example, if we prepare a large number of hydrogen atoms in the ground state, a measurement of the radial position of the electron will yield a distribution of values with an average equal to the radius of the simple Bohr model. But on occasion, the electron is found very close to the proton or several times further away than the average. The specific form of the statistical distribution can be calculated, for example, using Schrödinger's wave equation for the specific case of the hydrogen atom. Solutions of this differential equation are the so-called wavefunctions whose magnitude[15] at a given position in space is proportional to the probability of finding the electron there. Immediately after the measurement, the electron is located with certainty at the value found in the measurement. Therefore, the electron is no longer in the hydrogen ground state—the disruption has occurred. This change in state is somewhat dramatically, but also quite appropriately, referred to as the 'collapse of the wavefunction'. Measurements of other quantities, such as velocities, angular momenta, etc can be described in analogous fashion. Without a measurement, the system exists in all its potentialities. As soon as a measurement is made, one of the possible results materializes with a well-characterized probability. Erwin Schrödinger highlighted the strangeness of quantum measurements with an example he calls 'burlesque', which is probably not the adjective I would choose. A cat, by now widely known as Schrödinger's cat, is placed into a closed steel container together with a contraption that releases cyanide *if* a trigger is activated. The core of the trigger mechanism consists of a radioactive material with a dose chosen so that, over the span of one hour, a decay process, say the emission of an alpha-particle (see chapter 5), occurs with a 50:50 chance. If the decay occurs, the cat is dead, if it does not occur, the cat is alive. If no 'measurement' is performed during the hour, the cat is both dead and alive. As soon as the lid of the box is opened, a simple

---

[14] The 5th International Solvay Conference from 1927 in Brussels, Belgium, is a famous example. It brought together for the debate of the foundation of the new quantum physics 29 chemists and physicists, 19 of whom were or became Nobel Prize winners. Among the participants were (in alphabetical order) Niels Bohr, Max Born, Marie Curie, Lois de Broglie, Albert Einstein, Werner Heisenberg, Max Planck and Erwin Schrödinger.

[15] To be technical: wavefunctions can be complex valued (in the sense of complex as opposed to real numbers) so that their magnitude is the product of the wavefunction and the complex conjugate of the wavefunction.

inspection collapses this state to one of the two possibilities. This then, in a nutshell, is the Copenhagen interpretation.

Because of the probabilistic foundation of quantum physics, some of the early proponents, notably Albert Einstein, strongly objected and held out the possibility that the theory as currently formulated is incomplete. One way to cure nature of this severe case of gambling addiction was hoped to be a prescription of so-called hidden variables. Endowed with the right value, hidden variables were supposed to determine the outcome of any measurement process—no randomness here. However, the attempt to rescue any form of such realism in quantum physics has been dealt severe setbacks. In 1964, John Stewart Bell came up with a mathematical relation, now bearing his name, which can discern whether a system behaves according to probabilistic quantum rules or follows a path of determinism. Since Bell published his first paper, many experiments[16] have been performed to establish on which side of Bell's discriminating inequality nature lands. The unanimous verdict: nature does throw the dice and does so according to the rules quantum physics has found out. Today there is almost no loophole left to escape the fact that quantum physics is the correct and complete description of nature's behavior at the microscopic level. Furthermore, predictions of quantum physics have been verified experimentally with astonishing precision. For example, the calculated energy difference between certain states of hydrogen agrees with its measured counterpart to within one part in 15. A number with that many valid digits is analogous, for example, to a precision of the instantaneous Earth–Moon distance to within less than a hair's width—literally. So, as far as we currently know, quantum physics it is.

3.11 Fountain clocks—state-of-the-art time-keeping

No doubt other clocks will surpass and replace them, but for now cesium fountain clocks are the state of the art for time-keeping. Figure 3.6 already encapsulates the generic layout of atomic clocks, so that we can concentrate here on some details of the 'physics package'. From the generic EM spectrum in figure 3.5 and the energy–frequency relation for photons, $E = hf$, it follows that the energy splitting corresponding to the microwave transition used in Cs atomic clocks is about 100 000 times smaller than a typical separation of quantum states involved in the emission or absorption of visible light. The energy of the Cs transition [25] is also about 1000 times smaller than the kinetic energy of cesium atoms moving at room temperature. That this energy is so minute stems from the general nature of the splitting, which is an example of the so-called hyperfine structure of atoms. If you hold two bar magnets with like poles adjacent to each other (N next to N, S next to S), you will notice that this configuration is not stable. It tends to flip the magnets into the lower energy arrangement of opposite poles near each other. Although quantum physics changes the details, something very similar gives rise to the energy levels associated with the atomic hyperfine structure. Both valence

---

[16] The Wikipedia entry 'Bell test experiments' provides a fairly thorough summary of the current state of affairs.

electron of the cesium atom and its nucleus have magnetic properties that are very similar to those of a bar magnet. Because of the weakness of the magnetic interaction in general and the large distance—relatively speaking—the electron and nucleus keep on average, the energy difference between the two orientations of the magnetic dipoles is as small as observed and involves microwave radiation in the emissions and absorption process. The magnetic nature of the states is also the reason that emission of a photon is an unlikely occurrence, in other words the lifetime of the higher energy state is long. But we have seen that the Heisenberg uncertainty relation connects long lifetime with a narrow frequency spread. Which is a good thing for a stable clock and partly explains the high performance of cesium atomic clocks. The energy–time uncertainty relation holds in general. Therefore, the frequency uncertainty also increases with decreasing length of the interrogation time of the quantum oscillator. In other words, the longer you observe, the more consistent the clock frequency becomes. If an atomic clock could run with a single atom, and if that atom could be immobilized, we could probe the atom for an arbitrary long time. And if we keep the environment of the trapped atom stable, the clock will be extremely stable. While such clocks are in principle feasible, they do not currently exist as viable timekeepers. Cesium fountain clocks operate with about 10 million atoms at a time, but manage to keep them all going with nearly the same slow velocity that facilitates a long read-out time. After the introduction of the first fountain clock in 1999, NIST now has a second-generation clock (NIST-F2, item M in figure 3.1) online that loses a second in about 300 million years.

Briefly, the cesium fountain clock operates as follows. Cesium atoms are sent into a vacuum chamber, where they are cooled to almost absolute zero temperature at which all motion is frozen. Cleverly aimed and operated lasers accomplish the cooling by bombarding atoms with photons of just the right frequency. This slowing-down trick is a bit like stopping a bowling ball by throwing ping-pong balls at it. It can be done if you have enough ping-pong balls. Well, lasers emit tons of photons and so they can stop atoms in their track. As a result, the puff of cesium atoms contracts into a small ball, held in place by three pairs of laser beams traveling along the three spatial directions. When the laser pair along the vertical direction is momentarily adjusted to produce an imbalance of their respective photon flux, the ball of cesium is launched upwards. Once on their way, all lasers are turned off and the atoms, now subject only to gravity, follow a fountain-like trajectory that gives the clock its name. Just above the starting position, the tossed ball of cesium traverses a small microwave cavity on both its way up and its way down. If the microwaves have the right frequency f, they can induce transitions between the two hyperfine levels of cesium that are involved in the definition of the second. In other words, 9 192 631 770 cycles of the thus synchronized microwave define the duration of one second.

What is left to do is to figure out how to detect whether a transition has been induced, i.e. whether the radio station manager has found the correct frequency so that all atomic radio stations listen in. Yet another laser takes on that specific task. The frequency of this probe laser is selected such that it can only be absorbed by cesium atoms that have undergone the desired microwave transition. When they have, and only then, they can absorb a photon from the last laser—hitting them on their way down after they have crossed the cavity for the second time—which excites them to a much higher energy level. Subsequent light emission in the visible is the signal that all this has happened. Thus, adjusting the microwave frequency to maximize the detected visible light intensity guarantees that the former is indeed synchronized to the cesium hyperfine transition. Quite amazing, this tour de force to produce the best time standard. But it works!

3.12 A better clock—the ABC of clock comparison

You are not supposed to compare apples with oranges. I get it. Then again, it is easy to come up with all sorts of ways in which the two fruits differ or are similar—which might even explain why I like them both. In some ways, it is just as easy or difficult, just as fruitful (pardon the pun) or unproductive to compare apples with apples. Comparing clocks with clocks is no different. First, we should be clear which aspects we care about—price, quality, availability and/or consistency could be important factors for selecting one specific clock or apple over another. When we focus on quality, it seems much easier to characterize the excellence of clocks than that of apples. While many features—both objective and subjective—make up the value of the edible object, for clocks pretty much only precision and accuracy matter. But how do we know if a clock runs 'true'? What does that statement even mean? It simply means that the clock should indicate as faithfully as possible the chosen standard unit of time. Until 1960, the length of a 'mean solar day' served as our time unit, containing precisely $24 \times 60 \times 60 = 86\,400$ seconds. An average—the exact meaning was left to the astronomers—had to be incorporated into the definition to even out the seasonal and other variations of the length of a solar day. We also require a suitable device to produce the required 86 400 ticks per day, with each tick being worth one second. You can view Earth's rotation as the master oscillator and a mechanical or other type clock, if kept synchronized with the astronomical observations, as a faithful readout of the main timekeeper. Let us say we use for that purpose a well-adjusted pendulum, of the Rieffer–Shortt variety (see figure 3.1). If, over the long run, the pendulum clock ticks away the same number of seconds per day, all is in order. If not, we need to understand the origin of such a discrepancy. If we have reason to believe that the pendulum clock is flawed, we can tinker with the manmade device and let nature tell us when we succeeded to repair it. It says something about the quality of engineering when it turned out that we had to do it the other way around. Good old Earth's rotation was found to be erratic and we have adopted an approach in which we tinker with the length of the solar day by inserting leap seconds so that clock Earth can keep up with our atomic clocks. Since the latter were so much more stable, the second was redefined in 1967 so that the current version of the second is a certain multiple of the period of a certain cesium atom transition.

For better or worse, telling time is part of the modern phenomenon of globalization. The production and dissemination of precise time on a worldwide scale not only enables global traffic and trade, but telling time has become a global business itself. A network of several hundred atomic clocks, serviced by science and technology organizations in all continents but Antarctica, is the beating heart of International Atomic Time or TAI (from the French version, *Temps Atomique International*). In turn, TAI drives Coordinated Universal Time (UTC, an abbreviational hybrid between the English and French versions of the name, the latter being *Temps Universel Coordonné*), which is the time we use in our day-to-day lives. Apart from being an acronym challenge, UTC keeps track of time based on Earth's rotation, while TAI is counting seconds strictly according to the beat of the cesium atom of almost 10 billion Hertz. Not a single atomic clock, but the whole orchestra of atomic clocks mentioned above, is the embodiment of the definition of the second. Obviously, data exchange, communication and synchronization are technical hurdles that must be and have been cleared. If any new definition of the second based on a new type of clock—for example, one of the optical clocks currently under development—is going to supersede the present standard, it will have to demonstrate superior stability and precision in the same way as cesium atomic clocks proved their advantage over time-telling using Earth's rotation.

Ultimately, precisely defined metrics measuring the stability and accuracy of clocks decide which type of clock has an advantage over another. It is important to realize that the best performance of any clock cannot be had instantaneously. It takes time to know time. The degree of precision and accuracy with which a clock frequency can be produced is limited by random frequency fluctuations of the oscillator used. Such instabilities also depend on the duration over which the clock in question is being read out and averaged. Usually, operation improves with time, i.e. with the number of cycles of the cesium oscillations measured. This behavior is typical for noise reduction by averaging in general. However, for clocks, we eventually reach a plateau of precision and accuracy. Upon further increase of the interrogation time, clock performance deteriorates due to a combination of factors that depend on the specific circumstance and clock type. This generic behavior is captured in figure 3.11, which shows the Allan variance σ, a commonly used measure for deviations from clock perfection [26], as a function of read-out or averaging time τ. The smaller the Allan variance, the higher the degree of frequency stability. Well-characterized types of random fluctuations dominate various segments of the Allan variance, as shown schematically in the figure. Each noise type varies in a characteristic manner with frequency, as indicated in the figure caption. Only the so-called random walk noise (D) increases with sampling time and thus ultimately limits clock performance. For different types of clocks and even for different individual clocks of the same type, the minimum plateau in the Allan variance curve occurs at different read-out times. For some, best performance can be had more quickly than for others, even if the absolute minimum might be lower for the clock that 'needs more time' to perform at its best. In other words, the question 'which is the better clock?' may elicit the unhelpful but truthful answer 'it depends', because it depends on whether a quick response is important or whether you can afford to wait for the best performance. Currently, cesium fountain clocks have the lowest overall Allan variance, but it comes at the cost of having to average over prolonged periods of time.

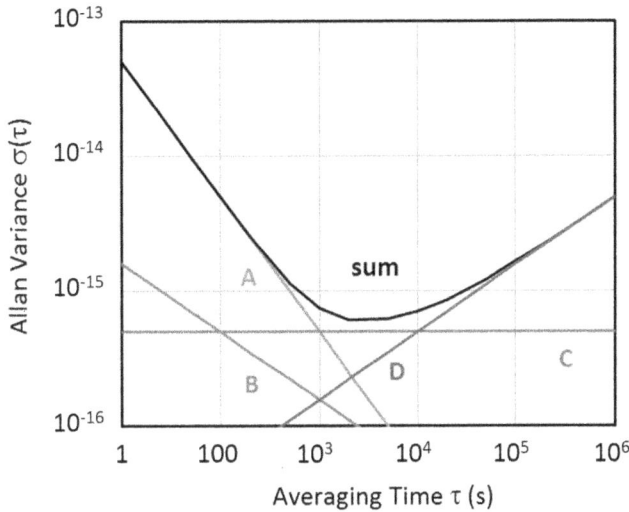

Figure 3.11. Generic Allan variance $\sigma(\tau)$ as a function of interrogation time. Dominant type of noise: A—$\sigma\sim 1/\tau$ (phase noise), B—$\sigma\sim 1/\sqrt{\tau}$ (thermal noise), C—$\sigma\sim$constant (flicker frequency noise), D—$\sigma\sim\sqrt{\tau}$ (random walk noise).

3.13 Who needs it?

It is one thing to improve the accuracy and precision of clocks so that they no longer lose or gain minutes in a day—Huygens did that long ago. But why would anyone care if a clock is off by a millisecond a day, let alone by thin time slivers lasting nanoseconds or less? Part of the answer lies in the fact that erroneous time-keeping accumulates to noticeable discrepancies. The Ptolemaic calendar builders found that out when they discovered the need for the leap day (see chapter 1). Earth's orbit around the Sun is about six hours longer than an integer multiple of Earth's rotation around its own axis. Thus, in four years, one 'extra' day accumulates and after 120 years things are out of order by a whole month. EM waves travel at a speed of roughly $3\cdot10^8$ m s^{-1}, about 30 cm per nanosecond[17]. Therefore, if an atomic clock onboard one of the global positioning system (GPS) satellites drifts one nanosecond per day, it appears to change distance to a fixed point on Earth—and vice versa—by about 30 cm in a day. Fountain clocks could do better (see figure 3.1), but they do not (yet) fit easily into satellites. Currently, atomic clocks onboard GPS satellites have a typical performance of about 100 ns uncertainty per day and thus must be synchronized and reset continuously. Yes, more compact, more precise clocks are useful.

In addition to its main purpose of finding position, the GPS provides reliable and precise time stamps for a variety of businesses [28]. Electrical power grid operators need to be able to direct large currents with high temporal precision, lest a local high-speed surge leads to runaway overloads of the grid and widespread blackouts—

---

[17] One of the few benefits of the length unit 'foot' used in the US: the speed of light is very nearly one foot per nanosecond.

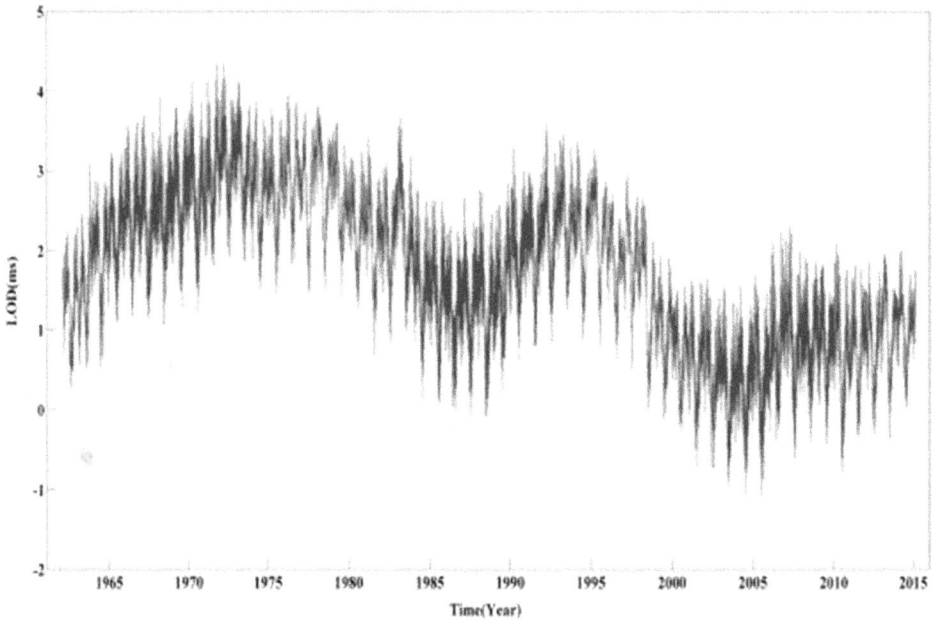

Figure 3.12. Length of day variation from 1962 to 2015. Data from compilation of the International Earth Rotation And Reference Systems Service [27].

as has happened on occasion. The organization of broadband communication and fast financial transactions are other examples. While these particular examples do not utilize the full temporal resolution of the atomic clock, they do benefit from the reliability and extent of the temporal web cast around the globe. Scientific research, on the other hand, has not only driven the enormous improvement in time measurement, it has also been a grateful customer of the highest precision that atomic clocks can offer. For example, the best available frequency standards are an essential tool in high-precision spectroscopy of atoms and molecules. Measuring the internal structure of the basic constituents of our everyday world in ever increasing detail is one of the most stringent tests of our theoretical understanding of nature. It also can lead to unforeseen applications, such as atom interferometry.

In geophysics, precise time-keeping tools have determined, as already mentioned above, the rate of Earth's spinning with nanosecond precision [29]. The data, now spanning more than four decades and reproduced in figure 3.12, show seasonal variations due to imbalances of the water circulation in the northern and southern hemispheres, long-term trends due to the influence of the Moon, as well as short-term and less regular variations that are not yet well understood but are possibly due to changes of processes in the interior of the Earth. Compared to Earth's irregular spin, pulsars, fast-spinning neutron stars, are the ultimate of consistency. These exotic objects were first detected by the astonishingly regularity with which they emit EM radiation pulses, a characteristic that earned the first ever discovered pulsar the nickname LGM, for 'little green men'. Rather than being an attempt to communicate by aliens, the pulsar signal is the result of the dynamics of a dead star of solar mass or so compressed into a sphere of tens of kilometers. The extreme decrease in size results in an enormous multiplication factor of the original, much more leisurely, spin rate of the star just before its death—an ice skater doing pirouettes on a gigantic scale. For example, a pulsar, discovered in 1985 at the Arecibo Observatory in Puerto Rico , has a period of about 1.5 ms, i.e. the star turns about 600 times per second. Since the precise period of 1.55 780 645 169 838 ms changes only very slowly, the spin rate of this and some other pulsars can rival atomic clocks in stability. However, we needed atomic clocks to establish these unusual pulsar properties in the first place. Here is one more example for the use of atomic clocks in astronomy: the anticipated observation of the black hole lurking at the center of our galaxy. Black holes, whose existence is implied by Einstein's theory of general relativity, originate from the gravitational collapse of stars a few times more massive than our Sun. The Event Horizon Telescope [30] is a network of radio telescopes scattered across the globe and strung together to become effectively a single instrument that is almost the size of the Earth. It is the stringing together part that is the key. Each telescope observes independently, but time stamps the radio data as they arrive with the help of synchronized atomic clocks. A central processing unit can then sort out features in the record of the different telescopes that came from the same source. Therefore, the reconstructed signal has the resolution of an Earth-sized telescope—enough to image the black hole against the bright radio background. As we will see in the next chapter, atomic clocks also featured prominently in confirmations of Einstein's special relativity theory—which comes next.

References

[1] http://www.nist.gov/pml/div688/2013_1_17_newera_atomicclocks_3.cfm
[2] Boucheron P H 1985 Just how good was the Shortt clock? *Bull. Nat. Assoc. Watch Clock Collectors* **27** 165–173
[3] Benedict F G and Lee R C 1936 The heart rate of the elephant *Proc. Am. Phil. Soc.* **76** 335–41
[4] Ossiander M *et al* 2016 Attosecond correlation dynamics *Nat. Phys.* **13** 280–5
[5] Ipp A, Keitel C H and Evers J 2009 Yoctosecond photon pulses from quark-gluon plasmas *Phys. Rev. Lett.* **103** 152301
[6] Rhode Island Historical Society *Cady Family Papers* www.rihs.org/mssinv/Mss326.htm
[7] Lombardi M A 2011 The evolution of time measurement, part 2: quartz clocks *IEEE Instrum. Meas. Mag.* **14** 41–8
[8] Maxwell J C 1868 On governors *Proc. Royal Soc.* **100** 1
[9] http://www.nist.gov/node/774256
[10] http://ieeemilestones.ethw.org/Milestone-Proposal:First_Atomic_Clock
[11] Codella C *et al* 1997 Four dense molecular cores in the Taurus Molecular Cloud (TMC)—Ammonia and cyanodiacetylene observations *Astron. Astrophys.* **324** 203–10
[12] Lawrence W L 1945 'Cosmic Pendulum' planned *New York Times* (21 January 1945)
[13] http://www.nobelprize.org/nobel_prizes/physics/laureates/1944/rabi-bio.html
[14] https://electronics.stackexchange.com/questions/187681/can-a-radio-transmitter-somehow-detect-the-number-of-receivers-in-its-area/187686

[15] Ahmed M *et al* 2008 The Brazilian time and frequency atomic standards program *Ann. Acad. Bras. Cienc* **80** 217–252

[16] *Work* Programme of the International Bureau of Weights and Measures for the four years 2016–2019 http://www.bipm.org/en/bipm/; Gibney E 2015 Atomic clocks face off *Nature* **522** 16–17

[17] von der Wense L *et al* 2016 Direct detection of the 229 Th nuclear clock transition *Nature* **533** 47

[18] Feynman R P 1967 *The Character of Physical Law* (Cambridge, MA: MIT Press)

[19] https://physics.nist.gov/cgi-bin/cuu/Category?view=html&Atomic+and+nuclear. x=82&Atomic+and+nuclear.y=13

[20] Bodner G M and Pardue H L 1994 *Chemistry, Study Guide: An Experimental Science* (New York: Wiley)

[21] Beyer A *et al* 2013 Precision spectroscopy of atomic hydrogen *J. Phys.: Conf. Ser.* **467** 012003

[22] Nobel Media AB 2014 Niels Bohr—facts http://www.nobelprize.org/nobel_prizes/physics/laureates/1922/bohr-facts.html

[23] NIST https://physics.nist.gov/cuu/Constants/index.html

[24] Peres A 2002 Popper's experiment and the Copenhagen interpretation *Stud. History Philos. Mod. Phys.* **33** 23 arXiv:quant-ph/9910078

[25] http://hyperphysics.phy-astr.gsu.edu/hbase/acloc.html

[26] Major F G 2014 *Quo Vadis: Evolution of Modern Navigation* (Springer)

[27] https://datacenter.iers.org/eop/-/somos/5Rgv/latest/214

[28] http://www.gps.gov/applications/timing

[29] McCarthy D D 2004 Precision time and the rotation of the Earth *Proc. Int. Astron. Union* **IAUC196** 80–197

[30] Homepage of the Event Horizon Telescope collaboration: http://eventhorizontelescope.org/

Chapter 4

Space and time forever entwined

By the middle of the second decade of the 21st century, only a few places on the land surface of the Earth remain uncharted. There are some unsettled issues though. Take, for example, the bragging right to be the tallest mountain in your country, Myanmar to be specific. In 2015, the *National Geographic* magazine ran a captivating story about an expedition trying to answer that question [1]. In the end, the quest failed because the team was unable to reach the summit of Hkakabo Razi. But why is it necessary in the first place to climb a mountain in order to measure its height? The answer is that it is not necessary at all. For quite a while already, mountain peaks have been sized up without scaling them. That is also the case with the mountaintop in question. As a matter of fact, its height had been determined more than once. And therein lies the problem, because the various data scatter so much that it is not clear whether a neighboring mountain might actually be taller. In the olden days, measurement of topological features was done exclusively by ground-based triangulation, which is nothing but an application of Euclidian geometry. Carefully determine a set of angles and at least one distance between points defining an array of connected triangles, and use the mathematical relationships between these quantities to find, say, how high a specific point is above sea level. The expedition had planned to do exactly the same thing, except using GPS. Standing on the top of Hkakabo Razi and using at least four satellites in reach as triangulation points, you obtain an answer that is usually more precise than results obtained the traditional way. Provided, of course, you manage to stand on top of the mountain.

One hundred years earlier and thousands of kilometers away, two key ingredients of GPS were discovered in another remarkable quest—this one successful. The ingredients stem from Einstein's special and general relativity theories, which were published in 1905 and 1915, respectively, and which predict that clocks run slower when they move relative to the ground, and that they tick faster when hoisted up from Earth's surface. Because GPS triangulation is accomplished by comparing the

doi:10.1088/978-1-6817-4096-6ch4   4-1

travel time of electromagnetic signals exchanged between a GPS receiver on the ground and satellites carrying synchronized atomic clocks, knowing the exact rate of these clocks is crucial. In this chapter we will find out what the special and general relativity theories contribute to our story about clocks and time. We will discuss the crucial aspect of measurement and try to illuminate the process of theory building, how from one idea and from bits of empirical evidence a fully fledged structure of space and time emerged that is starkly at odds with our everyday experience and yet has withstood all tests so far.

4.1 The ether and the birth of interferometry

One might say that the events leading up to Einstein's relativity theories included another unsuccessful endeavor. At least that is the way it appeared at the time. The goal was to detect Earth's motion through space by sensing differences of the speed with which light waves travel through their natural medium. Not a trace of that motion was found, though. That light is a wave phenomenon had been convincingly established many decades earlier. Because mechanical waves, which are nothing but an altered state of the medium they travel in, require a substance to undulate in or on—such as air for acoustic waves or a liquid for surface waves—it was natural to assume that light needed something analogous. That something was supposed to be the 'luminiferous ether'. This beautifully named essence was imagined to fill every bit of space, no matter how small or vast. After all, light traverses the space between Earth and any distant star, and it travels through material of all kinds of sizes and compositions. In that sense the ether could be seen as a manifestation of Newton's absolute space, 'filling', so to speak, space itself—space was the mold, ether the pudding. The properties of the reputedly all-permeating ether had to be extreme. On one hand, because Earth and all other planets do not show any sign of slowing down, the ether cannot exert any appreciable drag and thus must be very tenuous. On the other hand, it was also expected to be very stiff because of the enormous speed at which light travels through space (more on that below). The rationale for this latter expectation derived from the general behavior of mechanical waves, whose speed increases with the stiffness of the medium. For example, acoustic waves generated by earthquakes travel much faster through rock than through softer ground. Also, as we saw in chapter 2, a stiff spring bounces back more swiftly than a soft one. Therefore, although the ether was primarily invoked as the medium for electromagnetic waves, its fundamental role was clear and towards the end of the century it was a much discussed topic in physics. The big question was, how can we detect the ether? Since Copernicus, we have known that the Earth does not sit idly at the center of the Universe, but that it rounds the Sun once every year and turns on its axis once every day. Therefore, the motion of the Earth should give rise to an 'ether wind', with variable speed and direction. Maybe this ether wind can be measured somehow?

Just as sound waves in air or surface waves on water move more quickly between a stationary emitter and receiver when the medium itself moves from source to detector (and conversely more slowly when it moves in the opposite direction), it was fully expected that the travel time of light between two points in a laboratory on Earth would be affected by its motion relative to the ether. If we arrange a light wave to be split, say by a partially reflecting mirror, so that one half of the wave travels parallel to the ether wind and the other half travels perpendicular to it, then the two portions should move with different speeds relative to the laboratory. Then all that is left to do is to measure the difference in speed along the two paths. This proposition sounds and actually is reasonable. However, before we get our hopes up, we should have an idea about the magnitude of the two speeds involved. On its annual orbit around the Sun, the Earth moves along with a tangential velocity of about 30 km s^{-1}. The daily rotation around its own axis is associated with a mere 0.46 km s^{-1} at the equator and even less at higher latitudes. We can anticipate, then, that the order of magnitude of the ether wind speed, which we will denote by the label v, will be at most about 30 km s^{-1}. At around the time the ether concept was put forward, several experiments had measured the speed of light, customarily denoted by the letter c, and had found a value of around 300 000 km s^{-1}—about 10 000 times faster than Earth's orbital speed. This does not look promising to begin with and gets far worse. A more careful analysis—outlined in the next paragraph—reveals that the expected time differences are not of the order of the ratio of the two speeds, v/c, but only of the order of the ratio of their squares, v^2/c^2, about one part in 100 million. If you are of a more pedantic nature and happen to be concerned about your daily salt consumption, you might be inclined to follow recommended limits to the grain[1]. In that case, all you need is a paltry counting precision of the order of one grain per about 20 000. A precision of one part in 100 million is roughly equivalent of knowing the exact number of grains of a little more than seven kilograms of salt (or almost 16 pounds if you prefer non-metric units). If you count one grain every second without error, and do not sleep, eat, or do anything else, you will be done with the salt pile in about three years. Of course, counting grains of *anything* is not worth that kind of effort, but exploring nature's mysteries is a different story in my books—hats off to Michelson, Morley and all the others who did and do what most of us would not want to even start and probably could not accomplish (yours truly definitely included).

Maxwell [2], who had laid the theoretical foundation for electromagnetic waves (see chapter 3), was very interested in any experiment that—pardon the pun—could shed some light on the ether. He was also one of the first to point out the correct proportion between ether wind speed and expected time difference. In order to understand why this relation holds, we must analyze the times of travel for light propagating a distance L through the alleged ether. First imagine light moving in

---

[1] The internet is a wonderful thing. You don't know the average size of a grain of salt? Look, for example, here http://waynesword.palomar.edu/pinhead.htm. Multiply the grain volume with the density of NaCl (2.65 g cm^{-3}) to obtain the average mass of one grain and compare it to the recommended daily intake (e.g. 2.3 g per the US FDA). Since salt grains are not stacked in an orderly way, I include an estimated fill factor of 60% for loose random packing. And, yes, I found all of the necessary numbers on the internet, including the latter: https://en.wikipedia.org/wiki/Random_close_pack.

still ether with a speed denoted by c. When light moves against or with the ether wind, it should have a speed $c - v$ or $c + v$, respectively. Since, in general, the time needed to travel a distance L is given by the ratio of distance and speed, these times are here L/c, $L/(c - v)$ and $L/(v + c)$, respectively. Adding the last two values yields the total time for the light propagation in the segment of the set-up that is parallel to the ether wind. It is tempting, but would be wrong, to use $2 \cdot L/c$ as the round-trip time for the arm that is perpendicular to the ether wind. Rowing a boat across a river is not the same as doing the same across a lake, even if the two shores have the same beeline distance. Because of the current in the river you need to aim your boat upstream and the distance traveled *relative to the water* is longer for the river than for the lake. The same is true for the light propagating normal to the ether wind versus light traveling in the still ether. A careful mathematical analysis[2] of the situation leads to the conclusion that the fractional time difference between a round-trip across the ether wind and a forth-and-back trip of the same length parallel to the wind is equal to the square of the velocity ratio, i.e. $\Delta t/t = v^2/c^2$, as advertised above. Given the state of the art of experimental technology, Maxwell and most everyone else were highly skeptical that the necessary precision could be accomplished any time soon. For a while, the idea of the ether seemed to be an untestable idea—something that does not sit well with scientists in general and some experimental physicists in particular. Albert Michelson (1852–1931) was one of them.

Many textbooks of special relativity do not dwell on Michelson's experiments because they did not play an important role in Einstein's thinking and they are not a necessary ingredient for a conceptual introduction to the theory. While they are indeed dispensable, I want to provide some details to give a sense of the challenges Michelson and Morley were facing and to emphasize their impressive achievement. However, there is something else that motivates me to include a discussion of their experimental approach. The interferometric method they developed is the basis for stunning technological and scientific accomplishments that continue to this day—witness the first detection of gravitational waves by the team of the Laser Interferometer Gravitational-Wave Observatory (LIGO) in 2015. It is only fitting that this latest investigation of the fabric of space and time came about using the same basic approach Michelson invented over 130 years earlier. In 1907, he received the Nobel Prize in physics 'for his optical precision instruments and the spectro-scopic and metrological investigations carried out with their aid', as the official citation reads. This precision instrument is now known as a Michelson interfer-ometer and how it works we explain next. Because detailed discussion is indeed not required for what comes afterwards, you can skip the next three paragraphs and fast forward to the next section if technical details are not what you want right now.

The set-up sketch shown in figure 4.1 is taken directly from Michelson's and Morley's 1887 paper [3]. The entire optical arrangement is mounted on a massive

---

[2] This is a subtle point that even Michelson got wrong on the first attempt. Lorentz pointed out the mistake.

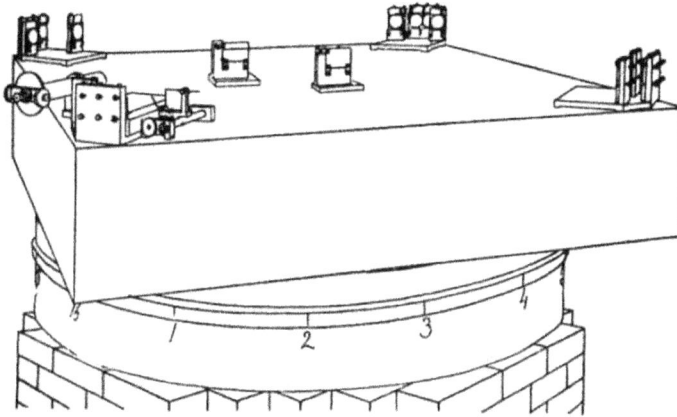

Figure 4.1. Perspective view of the interferometer used by Michelson and Morley in their 1887 experiment.

square stone that floats on a pool of liquid mercury (yes, rocks can float), which allows smooth, vibration-free rotation of the square block. At each corner sits a group of four mirrors[3]. Not shown is the light source mounted somewhere near the far left corner. A pencil-like light beam enters the interferometer via a focusing lens. At the front left of the stone you can see a viewing telescope. Finally, two adjustable glass plates are located near the center. One surface of one of the plates—the left one in figure 4.1 and the one labeled b in figure 4.2—has been coated with silver so that this surface acts as a partial mirror directing the light beam into the two separate paths. The second glass plate is needed so that the light rays along both paths of the interferometer are exposed to the same medium. That is it. No fancy electronics, and no computers. The optical and mechanical elements had to be of very high precision, however.

So far, I have given a description of the equipment. Here is what you do with it. Carefully align the light source so that a light beam shines onto one of the mirrors at the opposite corner. The beam will then also traverse the half-silvered mirror b (see figure 4.2) so that one portion of the light beam continues on its forward path, while the reflected portion reaches a mirror on the far right corner of the slab. The two light paths after separation by the beam-splitting mirror are commonly referred to as the two 'arms' of the interferometer. In the current case, the two arms are aligned along the two diagonals of the stone platform. Now comes the hard part. One by one and going back and forth, you have to align each of the 16 mirrors so that the light travels to a mirror diagonally across the platform. Once the light reaches the last mirror of each arm (labeled e and e' in figure 4.2), that mirror is fine-tuned to send the light exactly back onto itself. Therefore, the two returned portions will meet again where they split at the partially reflecting mirror. At that point, the light beams

---

[3] The purpose of using multiple mirrors on each corner is to increase—in this case by four—the distance over which the light beam probes the motion of the ether. The longer that distance is, the more sensitive the test comes out. Michelson and Morley could have used a larger stone, but supporting an even larger stone in a pool of mercury proved prohibitive. One of the breakthroughs in the LIGO experiment is the enormous increase in effective path length that light in the interferometer is allowed to travel before it recombines.

split once more, with one set of overlapping rays returning to the light source and one set being directed into the viewing telescope at the front left corner. In other words, with some patience and skill you finally manage to steer the light beam into the shape shown in figure 4.2. I told you it was hard, but the rest is not so bad.

What can you expect to see in the telescope where the original wave torn asunder at the splitting mirror is back together again? It is highly unlikely, even with the best alignment effort, that the two interferometer arms have exactly the same length and that therefore the two corresponding travel times are exactly equal. If, by chance, the travel times along the two interferometer paths differ by an amount it takes for one wave to advance (or fall behind) relative to the other by an integer multiple (0, 1, 2, etc) of the wavelength of the light, then two crests that started out together at the point of first split (at mirror M1) will be separated by that integer number of crests when they arrive at the telescope. In other words, while it is not necessarily the same crest they encounter, it is a crest (and one crest looks exactly like any other). Likewise, each trough meets a trough and the two waves are in perfect step—they recombine in what is called *constructive interference*. The result is a bright spot at the center of the telescope. Conversely, if the travel times differ by an amount so that the waves along the two paths slip by an integer plus one-half (0.5, 1.5, 2.5, etc), then a crest will meet up with a trough and vice versa. In that case, the two waves are

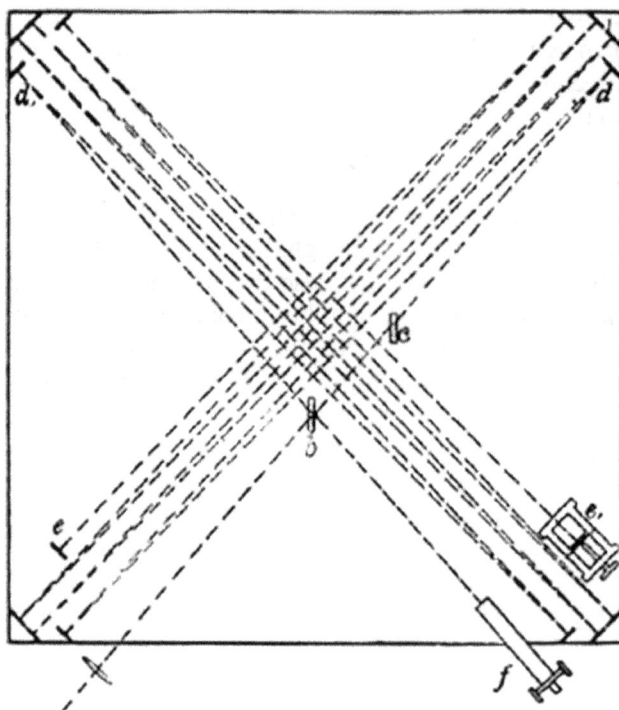

Figure 4.2. Schematic birds-eye view of the apparatus shown in figure 4.1. The dashed lines represent the light rays of the fully aligned interferometer. Note that the viewing direction in figure 4.1 is the right-to-left direction in figure 4.2.

perfectly out of step. When one is up, the other is down, and they cancel each other out: the intensity of the wave superposition is zero and a dark spot appears in the telescope crosshair. This condition is called *destructive interference*. Interference is a general behavior of waves; it is actually *the* litmus test of whether a phenomenon behaves like a wave or not.

Coming back to the operation of the Michelson–Morley interferometer, we note that one of the end mirrors (the one labeled e' near the telescope at the lower right corner in figure 4.2) can not only be tilted, but finely translated in the direction of the light beam in order to 'null' the distance difference between the two paths. As discussed above, at that moment crests fall on crests, causing a bright spot to be visible in the center of the telescope crosshair. If for whatever reason the length in one arm of the interferometer changes by as little as one-half of one wavelength, the interference becomes destructive and the spot turns dark. Because of the geometry of the waves entering the interferometer, conditions between constructive and destructive interference alternate as you move away from the exact center line in the telescope. A system of dark and bright rings (or depending on the mirror alignment) bands is visible. Such patterns of alternating bright and dark regions are known as interference fringes. For a given orientation of the instrument and at a given moment a specific, static pattern emerges. Now, we finally reap the benefit of having a floating interferometer. Give the stone a slight nudge so that it gently rotates around its vertical axis—maybe at a rate of a few minutes per 90 degrees of rotation. As the instrument circles, the alignment of its arms relative to the ether changes and with it the light speed in the interferometer arms. What was upwind becomes cross-wind and downwind etc. Thus, for a full rotation, the interference fringe pattern should shift periodically. For a difference change by one-half of a wavelength, the center bright fringe should turn dark and return to bright again after rotation by 90 degrees. A well-designed instrument can even reveal differences that are quite a bit smaller than one-half of a wavelength. And Michelson and Morley's instrument was indeed well designed and manufactured.

4.2 $c = 299\ 792\ 458$ m s^{-1} for everyone

Without question, Michelson and Morley would have been able to detect the motion of the ether had it been there. As hard and as often as they tried, they could not detect any fringe shift and thus any change in the speed of light. Michelson's first experimental test in 1881 in Berlin and Potsdam, Germany, was performed with a much cruder[4] apparatus and the absence of any shift of the intensity pattern was tantalizing but not yet convincing. Six years later, he and Morley implemented a version of the experiment in Cleveland, Ohio, that produced convincing results. The

---

[4] Nevertheless, the Berlin apparatus registered horse carriages driving by. This urban traffic disturbance prompted a move to the quieter suburb of Potsdam. Fast forward to LIGO. A staggering increase by many orders of magnitude in the capability to measure fringe shifts also implies a much higher sensitivity to background noise. Supposedly, the Hanford site instrument can pick up logging activity in the neighboring county. I would not be surprised if this superb capacity to measure Earth's motion will not simply be a nuisance but an interesting and possibly important aspect of LIGO, so to speak a fringe benefit.

null result was a shock. It was as if you threw a ball out of the window of a fast-moving train and yet the ball was not traveling any faster against the ground than if you had thrown it while standing still right next to the train tracks. A lot of smart people used a lot of clever tricks to try to reconcile Michelson and Morley's finding with the notion of the ether. If, for example, Earth was dragging the ether with it, then locally there was no relative motion and that could explain the negative outcome of the Michelson–Morley experiment. Nice try, but that kind of dragging is inconsistent with stellar aberration, a phenomenon with which astronomers were already quite familiar in the 19th century. Viewing a star at a fixed time of night over the course of a year requires that the telescope be swung by a small (±20.5 arc seconds) angle about a central direction [4]. An analogy for this is the challenge of catching rain drops in a tall and narrow cylinder in such a way that the drops do not touch the inside walls. If the rain comes down vertically, all you must do is to hold the container vertical. And if wind forces the drops to fall at an angle, you can tilt the cylinder appropriately. The same situation arises in case the rain comes down vertically, but you are running while catching the drops. Clearly, during the time that drops fall towards the bottom, the tube moves laterally and thus the drops may strike the wall. Tilting the container will again do the trick. This is exactly what happens in stellar aberration. The fact that the telescope tubes have to be tilted to catch the light raining down from the stars proves that there is lateral motion between the observer and the light rays. If the ether were dragged along by Earth, no such motion would exist.

Since dragging the ether does not work, maybe a more radical explanation of the Michelson–Morley experiment is warranted. Hendrik Antoon Lorentz (1853–1928) was happy to have a go at it. Suppose that at one moment Earth moves relative to the ether with some non-zero speed and that the speed of light in the laboratory is different along the path of motion compared to a direction that is perpendicular. If somehow all distances in the direction of motion are altered, the light travel times along the two arms of the interferometer might still be the same. Lorentz proposed a specific relation between length values and speed, which we will discuss later. Suffice it to say here that this assumption would indeed explain the Michelson null-result.

Unorthodox as Lorentz' proposal was, Einstein called for an even more drastic approach: there is no change of light speed because light moves at a universal, constant speed, no matter what, and there is no possibility of detecting absolute motion—none. However, in that case, not only spatial dimensions but time itself has to warp. From the constancy of the speed of light, everything else follows, including time dilation, length contraction and all that jazz. In his book *Relativity Theory Visualized* [5], L C Epstein gives an amusing and quite apt analogy for just how crazy Einstein's approach was. Imagine that a door in your house rubs against the floor and no longer opens and closes smoothly (aka the null result of the Michelson–Morley experiment chafing against the accepted physics framework). The remedy you and I would probably apply is to plane off the offending part of the door. Einstein's approach is to jack up the house and tilt its foundation (aka space and time as the basis of all of physics). Ordinarily, this is *not* a strategy one would recommend. But in this instance a sweeping fix was necessary—and in Einstein's hands it did work.

Suppose that, once the foundation has been changed, the creaky door is mended. But what about all the other doors in the house? While they operated just fine on the old base, now *they* might have become sticky. Before we can address that question, we need to examine in more detail the repair Einstein applied. As already stated, no change of the speed of light was detected in the Michelson–Morley experiments, because there is none. The speed of light is exactly the same for *any* observer, regardless of the state of motion of the observer or the light source. So no matter how fast you run when hurling photons—the grains of light—around you, no matter which direction you run in, the specks of light will always speed away from you at the same rate. And no matter how fast you run, any photon coming your way will be approaching at a rate of about 300 000 km per second. More precisely, the speed of light is $c = 299\ 792\ 458\ \mathrm{m\ s}^{-1}$ for everyone. Because of its universal consistency, the speed of light has taken a special role in our construction of units and universal constants. In 1983, the Conférence Générale des Poids et Mesures (CGPM) defined the length unit as 'the length of the path traveled by light in vacuum during a time interval of 1/299 792 458 of a second', thus establishing the vacuum speed of light to be exactly equal to these nine digits—no more, no less.

The absolute constancy of the speed of light is in direct conflict with our intuition about how velocities should add. Simply put, one plus one is no longer two but less than that—sometimes only a little and sometimes a lot less. This, of course, is already odd—and things are going to get more interesting. After all, the concept of velocity is related as directly as it gets to the twin concepts of space and time: the average velocity is the ratio of distance traveled and time needed for that travel, which in the limit of vanishingly short distances and/or time intervals leads to the definition of instantaneous velocity. If adding velocities is what we tinker with, it should not come as a surprise that notions of time and space will be radically altered.

Today, the hypothesis of an unchanging speed of light is supported by a large number of laboratory experiments as well as astronomical observations. For example, the light emitted by decaying elementary particles which themselves move at almost the speed of light is found to have the same speed as light emitted from those particles at rest. In the following section, we will therefore assume that the speed of light is constant. That simple principle will lead us to extraordinary conclusions.

4.3 The principle of relativity

Galileo was very clear about it. When locked inside a windowless cabin, you cannot tell if the ship on calm waters is moving uniformly against the shore or is sitting perfectly still. No tossing of balls, no hanging of a plumb line, no rolling objects on inclined planes—no mechanical experiment whatsoever—will reveal the difference between those states of motion. This in a nutshell is the Galilean principle. Accelerated motion is another story. Our sense organs will give us clear signs of the presence or absence of acceleration. And when rough seas toss the ship up and down, an object lying on the cabin floor presses harder on it or jumps up from it. With a strong breeze in the sails and sustained horizontal acceleration, the path of a dropped ball will no longer be vertical and neither will a plumb line. But constant, straight-line motion, Galileo says, cannot be detected by mechanical means.

As we saw in the previous section, 19th-century physicists thought that the motion of electromagnetic waves would violate the Galilean principle. However, the experiments by Michelson and Morley were incompatible with that idea. Einstein then reaffirmed the Galilean principle and placed at the heart of his theory of relativity an extended version, namely the impossibility of detecting absolute uniform motion by *any* means. Expressing the same idea more positively: special relativity starts with the premise that the laws of physics are precisely the same and certain universal constants have identical values for any observers, provided they either move at uniform speed relative to each other or not at all. General relativity adds the twist that acceleration—while still detectable as such—is indistinguishable from the effect of gravitational forces, and that effects connected to gravitational interaction can be described entirely by free motion along curved space–time. In other words, it is impossible to decide from within a closed box whether this box is being accelerated by an external agent or whether it rests on the surface of a large object with sufficient mass to generate the equivalent 'gravitational' acceleration.

Special and general relativity expand on a trend that had started with the Copernican revolution, of removing Earth from its unique place at the center of the Universe—now, there are no 'central' or 'special' places left to be found *anywhere*. Each inertial reference frame is as valid as the other and Newton's absolute space and time have disappeared. However, this does not imply a free-for-all universe. For example, the speed of light is now the same everywhere and for everyone. In addition, we will see that, despite time dilation and length contraction, a generalized 'distance' between any two events in space and time exists whose value is also universal and independent of the choice of reference frame.

In the next sections, we will explore the consequences of the relativity principle. For the most part, they turn out to be counterintuitive and puzzling. Not the first one, though.

4.4 A first (boring) application

Our intuition, based on experience with speeds very much slower than the speed of light, leads us to expect that the physical dimensions of an object stay the same regardless of the state of motion of the object. Actually, these are two separate expectations for unchanged size along and perpendicular to the direction of motion of the object. Indeed, any change of dimension along a direction *transverse* to the motion is also incompatible with Einstein's relativity principle. If moving objects were to shrink or expand along transverse directions we would end up with the following conundrum. Consider two coaxial, thin-walled cylinders whose diameters are exactly the same when both are at rest. Suppose the cylinder diameter were to contract if moving in the axial direction. Let A be the cylinder at rest and let B be the one that moves. Since B contracts, it will fit inside cylinder A when it moves towards

A. Afterwards, an objective investigation can unambiguously establish which cylinder was outside and which one was inside. For example, both cylinders could be equipped with flexible wires on their outside (think of a hairbrush). Thus the cylinder on the outside will have scratches on its inside, while the interior wall of the inside cylinder remains unblemished. In the above scenario, cylinder *A* will have scratches caused by cylinder *B* slipping through its interior. Obviously, cylinder *B* will emerge unscathed. From the point of view of cylinder *B*, cylinder *A* is moving. If the relativity principle is to be valid and uniform motion is undetectable, then it must be the case that cylinder *A* is now shrinking in diameter and thus causing the interior of cylinder *B* to be scratched. Switching points of views does *not* mean, though, that the event occurs twice. There is only one incident of moving cylinders and at most one cylinder can be scratched. But if only one cylinder were to be scratched, it would single out a 'privileged' reference frame, thus making motion absolute and detectable. The argument is the same if we assume that moving cylinders expand. There are many variations on this theme. For example, moving trains derail on their stationary tracks with their wheels ending either inside or outside of the tracks, while they should do the opposite when things are considered from the point of view of the train (so that now the tracks are moving). All such scenarios end the same way: the only way to keep the relativity principle alive in this first application is to require that *no* change of lateral size occurs when objects move. This is sensible enough and just as we had anticipated. But do not get your hopes up too much: embrace the notion that the speed of light is truly constant and weird things are bound to happen.

4.5 Moving clocks must run slow

Take, for example, two clocks, both of identical make and build, and running as smoothly and accurately as you can wish. Motionless side by side, they show the same time. But if one clock moves, it ticks more slowly. Halt the moving clock and they again tick at the same rate. Now move the other clock and, again, the moving clock runs slow. Odd, is it not? Even with a very expensive watch you will not be able to detect that driving in a fast car—or even flying in an aircraft—changes the time told by that watch. You have to invest quite a bit more money into a high-precision timepiece to be able to measure the predicted effect. But it is there. As already emphasized on previous occasions in this book, a conceptual understanding does not give you the full picture. Numbers are important to appreciate under what circumstances a conclusion drawn from theory has practical significance. In the next two paragraphs we will do both. As for the numbers, the enormous magnitude of the speed of light compared to most speeds of ordinary (aka human) experience ensures that relativistic effects are for the most part insignificant in everyday life.

Einstein was a master at conceiving *thought experiments* (or *Gedankenexperimente* in his native German). In such experiments, no pesky limitations due to poor materials, design, or funding exist—as long as the laws of nature are respected. And because you can change which laws you want to explore just as easily as the virtual equipment, thought experiments can be a powerful tool. Our aim is to investigate identical clocks in uniform motion relative to each other. Unrestrained by fiscal prudence, we pick the finest clocks available, which—elegantly simple—function without any moving mechanical parts. In an evacuated glass tube, a light pulse bounces back and forth between two perfect mirrors a distance L apart. At time $t = 0$, a finely tuned mechanism that is embedded in a mirror at one end of the tube emits a short light pulse that then races at the speed of light c to the opposite end. There, the pulse is reflected back towards the source end of the tube. Such a device is completely analogous to the mechanical clock model introduced in chapter 2 that was based on a particle bouncing inside a box. With light we no longer have to be concerned about the mechanism slowing down. Completing the new clock are sensitive detectors that are connected to fast and accurate counting electronics. Remember, we are not held back by sloppy engineering or manufacturing, and all components can be expected to work perfectly. If the circuitry is designed to count each mirror bounce as one temporal unit, then the travel time between two mirrors $T = L/c$ is that unit[5]. For example, a length of about 30 cm will yield a unit of one billionth of a second, i.e. $T = 1$ ns. With a bouncing light pulse of very short duration, say 1 ps = 0.001 ns, the spatial length of the pulse is much shorter than the clock length and our tool resembles the bouncing bead clock of chapter 2 even more. Figure 4.3 illustrates this device by a sequence of five different snapshots (or movie frames if you like). When each snapshot is taken a time $T/2$ after the previous one, then at the third frame the clock has just finished one 'tick', the fifth frame captures the second 'tick' and so forth. So far, so simple.

What do we find when the clock is moving? figure 4.3 can also be interpreted as a movie of exactly that situation. Specifically, consider a clock moving to the right at a constant speed V. Then frames 1–5 show not only the passing of time, but also the displacement in space of the moving clock. From the stationary camera point of view, the light pulse travels along the inverted V-shaped path outlined by the dashed lines in the figure. The speed of light is unchanged, but the distance traveled has increased. That can only mean that, observed from the ground, the moving clock

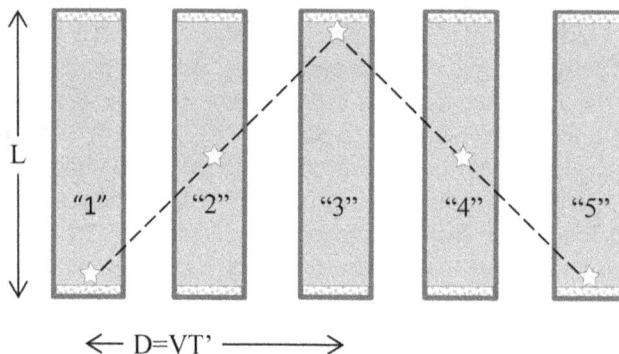

$$\longleftarrow D = VT' \longrightarrow$$

Figure 4.3. Photon clock. See text for how this clock indicates time.

---

[5] In section 2 of chapter 2, I argued that at least three ingredients are necessary for a proper clock: oscillator, counter and energy source. The oscillator is, of course, the reflected light pulse, counting is accomplished by the embedded detectors and any energy needed to compensate for absorption and/or scattering can be generated by appropriate design of the 'smart' mirrors.

produces a time unit T' that is *larger* than the corresponding unit T of a stationary one: *the moving clock runs more slowly than its stationary identical twin*. If you have a lot of questions and complaints about right now—great! In the brief discussion above, a few things have been swept under the rug to allow a smoother flow of the argument. In the end, though, the argument *does* hold. Before inspecting possible objections, I want to take the line of reasoning one step further.

We can use the above thought experiment to derive a quantitative prediction. We labeled the period of the moving clock T' (the time it takes for the light pulse to travel from one mirror to the next). Between frames 1 and 3 one of these time units has elapsed *and* the clock has moved a distance $D = VT'$ (the horizontal length between frames 1 and 3). Note that the time interval between two camera frames is no longer $T/2$, but a bit longer to accommodate the extra distance the light pulse has to travel, namely the distance L' (the length of the dashed line connecting the bottom of frame 1 and the top of frame 3). Because of the agreed upon definition of the unit of time, we are forced to identify L' with the distance light travels in the unit time T', i.e. $L' = cT'$. With a little help from Pythagoras, we find that

$$(L')^2 = L^2 + D^2 \Rightarrow (cT')^2 = L^2 + (VT')^2.$$

Solving this equation for the time unit T' of the moving clock yields

$$T' = \frac{L}{\sqrt{c^2 - v^2}} = \frac{\dfrac{L}{c}}{\sqrt{1 - \left(\dfrac{v}{c}\right)^2}} = \gamma T, \quad \text{with} \quad \gamma = \frac{1}{\sqrt{1 - \left(\dfrac{v}{c}\right)^2}}, \tag{4.1}$$

where I have used the fact that $L/c = T$. The relation expresses moving clock time T' in terms of stationary clock time T. As we will see in a moment, for any non-zero speed the time indicated by the moving clock is always larger than that of the stationary clock—an effect aptly called time dilation.

At this point, using the words 'time dilation' might appear to be somewhat of a leap of faith. After all, the above line of reasoning is only based on arguments concerning specific clocks—imaginary ones at that—in specific circumstances. Nevertheless, in the framework of special relativity the behavior deduced from the thought experiment is not indicative of a faulty clock; it does reveal an intrinsic, not very intuitive property of what we call time and motion. We are in the process of exploring the consequences of Einstein's theory—not proving or disproving it. Therefore, we need to push the premise of the principle of relativity, and particularly the assumption of the constancy of the speed of light, as far as we can. Imagine, then, that we complement the light clocks, both stationary and moving, with an assortment of other items that change with time: a quartz watch, an atomic clock, a petri dish with dividing yeast cell cultures, a box with radioactive material, a bottle with chemical compounds that slowly react with each other and a monitor logging the heart beats of identical twins. You can continue the list at will. If the principle of relativity is to be valid, all the ways of telling time must agree so as not to single out one reference frame over the other. Heart beats, atomic oscillations—everything slows down, not just the ticks of the photon clock. Special relativity really does predict time dilation. Of course, in the end only experiments can render a verdict. Spoiler alert: so far all observations have been consistent with the theory.

The above-defined quantity γ is not just a convenient shorthand, it is an important and ubiquitous scaling factor between stationary and moving frames in special relativity. From the definition of γ we can see that in the case of two stationary frames, i.e. for $V = 0$, the factor is unity ($\gamma = 1$), which is as it should be: no change between identical clocks under identical conditions. However, if V approaches and eventually matches the speed of light, the denominator decreases and for $V = c$ becomes zero, in other words γ grows beyond all bounds as V approaches the speed of light. One second stretches into a minute, a year, infinity—a clock running at the speed of light stands still and photons do not age. Speed values beyond c render the radicand negative. Hence γ becomes a complex number, signaling non-physical behavior. The speed of light acts like a mathematical barrier here and indicates an exclusion of speeds higher than c. Arguments based on causality bolster this prohibition of faster than light motion for real objects.

Numerical evaluation helps to determine the situations in which matters of principle are also matters of practical importance. Using the time dilation equation (4.1), we can find out what the fastest mode of transportation would do to your watch. Which brings up the interesting question of what that mode of transportation is. Any ideas? Yes, rockets are by far the quickest way to move. The Apollo 10 mission for example reached a peak speed of almost 40 000 km h^{-1} relative to Earth. If you are willing to dispense with human passengers and also accept velocities relative to the Sun, then the fastest engineered object may have been the Helios satellite[6], which in its solar orbit exceeded speeds of 250 000 km h^{-1} or ~69 000 m s^{-1}. Even this latter value is only a tiny fraction of the speed of light ($V/c = 2.3 \cdot 10^{-4}$) with a concurrent γ-value of 1.000 000 027, which is for all practical purposes equal to one. In order to gauge how tiny the relativistic correction is even for this pretty impressive speed, look at the fractional change between a clock on board Helios and one at rest relative to the Sun[7] i.e. $\Delta T/T = (T' - T)/T = \gamma - 1 = 2.7 \cdot 10^{-8}$. This small number implies that every day a difference between the clocks of only a little more than two milliseconds accrues. Riding a much, much slower plane you clearly need ultraprecise atomic clocks to tell that time is slowed—but then you *can*.

We have some unfinished business to attend to—we need to clean the 'dirty' details from under the carpet and make sure they do not invalidate the arguments leading to the time dilation equation. For starters, concerns could be raised that the thought experiment does not sufficiently conform, as required, to all physical laws.

---

[6] Here is another contender for fastest manmade object. On 4 July 2016, the Juno probe was at its peak velocity of about 100 000 km h^{-1} relative to Earth before a braking maneuver provided sufficient deceleration to inject the satellite into orbit around Jupiter—another stunning proof that we understand celestial mechanics pretty well.

[7] We cannot place the reference clock on the Sun itself, so we must be more creative. While we do not want our clock to melt, that issue is *not* the reason to avoid the Sun—we can count on our limitless engineering capacities when performing thought experiments. But we still have to obey the laws of nature, and they include an unavoidable differential effect of gravity on clocks located at the Sun and those far away.

For example, real light pulses have a finite duration, spread out in the lateral direction and impart momentum to the mirrors. Maybe such effects could undo the deduced time dilation? In most cases, these objections are not at all fundamental and can be addressed by proper engineering (for example, making the mirrors sufficiently massive). More importantly, time dilation is already present after one period of the clock. Therefore, the demands on precision or energy consumption are finite and do not play a role in the thought experiment. However, one can come up with at least two basic issues (I am not sure if there are more) that cannot be resolved so easily. Interestingly, in trying to answer these reservations we will be led even deeper into the maze of special relativity.

We have already seen that Lorentz proposed length contraction to explain the null result of the Michelson–Morley experiment. Therefore, it is conceivable that a lateral contraction compensates the longitudinal distance traveled by the moving clock so that the light paths in the stationary and the moving clock are of identical length after all. Then no time dilation would occur. But we have seen in the previous section that this is not the case. Another issue has to do with the asymmetry in how the ticking of stationary and moving clocks is measured. In the first case, we need two clocks side by side—one our time standard, and the other to be investigated. In the latter case, we really need at least three clocks—the moving clock under investigation and *two* standard clocks, while these two clocks need to be properly synchronized. Maybe this aspect gets in the way of our argument? The following section will explain why we do not have to worry about this possibility, either, by clarifying the central role of proper clock synchronization. After that, we will be equipped to tackle the funny business of length contraction.

4.6 When lightning strikes twice—was it or was it not simultaneous?

Why is it not possible to rely on a single clock, say located at the starting point of the moving clock (frame 1 in figure 4.3) when we want to measure the rate of a moving clock? Here is the answer in Einstein's own words (my translation from his 1905 paper 'On the electrodynamics of moving bodies'—his special relativity manuscript): 'If we want to describe the motion of a material point, we specify its spatial coordinates as a function of time. It is important to keep in mind that such a mathematical description only makes physical sense if we are clear what is meant here by "time". We need to take into account that all our declarations in which time plays a role are declarations about *simultaneous events*. For example, when I say "This train arrives here at 7 am" then this means roughly "The pointing of the hour hand of my watch to the number 7 and the arrival of the train are simultaneous events."' Einstein emphasizes that this definition of assigning time to events at one location—as cumbersome as it may appear—is necessary, unique and sufficient. However, when events occur at two different locations, the need arises to compare two separate clocks.

For two events at separate locations to be classified as simultaneous, the two local clocks must read the same time when the respective event occurs *and* the clocks must be properly synchronized. It does no good to make an appointment when one of the participants uses a clock that runs late. Here is a simple synchronization method that is consistent with the guiding principles of special relativity. Arrange a certain number of identical clocks on a square grid in two or three dimensions as needed. Choose the location of one of these clocks as the origin of a spatial coordinate system. The clock at the origin serves as the 'reference' (or 'primary' or 'standard') clock. All other clocks are referred to as 'secondary' clocks. Let L_n be the distance of secondary clock number $n = 1,2...$ from the primary clock[8]. At time $t = 0$, let a flash of light be emitted at the location of the primary clock. This flash is of negligible duration and spreads out in all directions. Immediately upon arrival of the light pulse at clock n, set that clock to read $t_n = L_n/c$. All clocks in the network are then properly synchronized—at least in principle, which is all we need for our following thought experiments. To be sure, clock synchronization across real networks such as the internet or between clocks on board of satellites is not a trivial task, in part because light changes speed when it travels in any other medium than the vacuum. An implicit assumption made in the synchronization procedure described above is the isotropy of the speed of light in empty space. In other words, we assume that vacuum speed of light is not only constant along a given direction, but that it is also *the same in any direction* in space. We have no reason to think that a symmetry-breaking directionality exists in empty space, but we should be clear about this point (as Einstein was in his 1905 paper). In any case, we can now meaningfully compare clock readings in a given reference frame.

Suppose, for example, that a large number of synchronized stationary clocks has been placed along the path of motion in figure 4.3. Suitable local sensors can then record both the state of the moving clock (i.e. the exact location of the light pulse) and the time at which that state occurs in the stationary frame. At the end of the experiment, all such readings can be combined at leisure into a well-defined history of the moving clock. Every single entry in the log comes from a properly synchronized local clock providing a valid time stamp in the spirit of Einstein's idea. Such carefully controlled data acquisition allows us to proclaim unambiguously that the period of a moving clock is subject to the time dilation equation derived above.

Now that we understand what is meant by finding two events to occur simultaneously, we will see that according to special relativity simultaneity cannot be absolute[9]. Observers in two different inertial reference frames will agree on simultaneity when the two events in question occur at one single location (otherwise all sorts of illogical consequences would occur). However, the observers can and often will disagree on the temporal interval between events at two separate locations. Under certain circumstances, it is even possible that the order is reversed: an event B

---

[8] The distance between point A and point B is obtained by laying standard rulers in a straight line between A and B and counting how many rulers are necessary. Fractional ruler lengths are allowed down to the desired precision. Nothing is moving in this procedure (take as much time as you want), so we can safely assume that the measurement of length is well defined.

[9] We are still in the process of exploring the consequences of Einstein's theory—not proving or disproving it. In the end, only experiments can render such a verdict. And even experiments can only falsify a theory by finding discrepancies. Finding agreement only means that observations are consistent with the theory so far.

Figure 4.4. Early prototype of a mobile instrumentation platform.

that happens *after* event *A* in one frame can occur *before* event *A* in another reference frame.

Lightning supposedly does not strike twice—in the same place. Obviously, that does not apply to the case of two bolts in two places. Imagine, then, two spots a certain distance apart being struck by two lightning bolts. How would you know if the two events happened simultaneously? With the above outlined definition for synchronizing clocks we know what to do. We blanket the predicted thunderstorm area with a network of synchronized clocks that are designed to stop at the moment a lightning bolt hits within a short distance Δx. After the storm has subsided, we go out into the field to collect our data. Wherever we find a stopped clock, we mark time and position. Indeed, we discover a pair of clocks, *A* and *B*, which are a distance $L \pm \Delta L$ apart and have stopped at the same time $t \pm \Delta t$. Here, the quantities Δx, ΔL and Δt denote inescapable measurement uncertainties in real experiments. In a thought experiment we can assume that these limitations shrink to zero. In other words, the two lightning bolts at *A* and *B* have hit the ground simultaneously.

Now comes the interesting part of our new thought experiment, the part that really demands unlimited funding and superb engineering. While the storm is raging, we send out our highly sophisticated mobile instrumentation platform (MIP). MIP is a large flat scaffold that hovers a short distance above the ground so as not to disturb the stationary clocks below. The platform is made of a material that is transparent to lightning bolts yet strong enough to carry its own network of synchronized clocks and light sources. Think of it as a high-tech version of Prince Husain's famed magic carpet in one of the tales of *One Thousand and One Nights* (or see figure 4.4). The clocks are identical to those on the ground and their synchronization has been performed in an identical manner after MIP was launched and reached its final cruising speed *V* relative to ground. As luck would have it, the platform motion was exactly in the direction from *A* to *B*, the same points where the two lightning strikes hit the ground. After the storm is over, we stop MIP and read out its data bank. What do we find? We find that the two MIP clocks recording the lightning bolts in question did *not* stop simultaneously, specifically, the MIP clock located at event *B* stopped *after* the MIP clock recording event *A*. Everybody followed protocol to the letter and all of the clocks functioned perfectly. So what went wrong? Of course, standing on the ground we notice that all of the MIP clocks are running slow, as discussed in the previous section. But that would only account for discrepancies in time intervals between asynchronous events. Time dilation alone does not explain the lack of simultaneity. Something else must have happened.

The fundamental reason for the discrepancy is a disagreement on whether proper synchronization has been achieved—despite the fact that the same method was used. This sounds bizarre, but it is an inevitable conclusion if we accept the validity of the principle of relativity. Here is what ground clocks detect when the synchronization procedure is carried out on MIP. Since motion is along the direction from *A* to *B*, we can concentrate on ground clocks along that line. At one point, synchronization is initiated on the MIP by briefly flashing its central reference clock. Subsequently, the light pulse spreads in all directions, including along our line of interest, which is also the direction of relative motion between the MIP and ground. On the moving platform itself, the flash reaches pairs of clocks simultaneously that are equidistant from the MIP origin. However, on the ground things play out differently. Because the speed of light is finite and constant, the sync flash takes longer to reach that clock of a given pair that is located along the forward direction (*A* to *B*), since that clock is moving away from the flash. Conversely, the clock of the pair in the backward direction (*B* to *A*) will be reached earlier because it is moving towards the flash. In other words, as determined from the ground, two clocks equidistant from the MIP origin are *not* synchronized by the procedure that *does* synchronize the clocks as determined on MIP itself! Of course, two unsynchronized clocks will measure two simultaneous events as occurring at different times. So no wonder that the MIP clocks disagree with the verdict obtained on the ground that the two lightning bolts struck *A* and *B* simultaneously. By the way, the situation is completely symmetric. From the MIP point of view, the clock network on the ground is improperly synchronized, thus explaining the observed discrepancy. The lack of simultaneity between the two bolts on MIP is just as real as the existence thereof on the ground. Here again, we are not dealing with faulty procedures or broken equipment. Everything works as it should. Variance in the assessment and reality of simultaneity is 'simply' an aspect of the way time unfolds in reference frames moving relative to each other.

A comment is in order about how classical, Newtonian mechanics describes the above situation. In the context of pre-relativistic physics, simultaneity or the lack thereof is absolute. From the detailed description above it is clear that such an absoluteness must have a price. That price is the way speeds add. In the above scenario of the MIP synchronization, the lack of simultaneity came about because the speed of light is constant no matter what. In classical physics this would not happen because the speed of the light flash in the forward direction is the sum of the light speed against the ether and the speed of the MIP against ground. For the flash in the opposite direction, the two speeds are subtracted. Therefore, the two clocks are also reached simultaneously in the MIP procedure. Unfortunately (for the Newtonian world view), Michelson and Morley put an end to this possibility. The speed of light does not change, there is no ether wind, one plus one is not two—the constancy of the speed of light is uncanny but real. Lack of agreement on issues of simultaneity can have the drastic result that the temporal order of events is reversed for observers in different inertial referent frames. In such cases, special relativity preserves the notion of causality—past and future can only be reversed if no causal relationship exists between events. In a very specific way special relativity even provides an *absolute* measure for the distance between two events in the form of the space–time interval. But before we get to discuss this bit of absoluteness, we need to discuss one more weird aspect of change, the so-called length contraction.

4.7 Running makes you thinner

There is a straightforward argument to make it at least plausible that moving rulers must shrink if the principle of relativity is true. Suppose the speed of light is measured by two observers with the help of rulers and properly synchronized clocks. At time zero a brief flash of light is emitted at one end of the one-meter ruler. Everyone agrees on the speed at which light races towards the other end of the ruler, namely 299 792 458 m s^{-1}—no matter what the observer's state of motion might be. Now let one of the participants move relative to the other with constant speed V. Nothing is different in the moving system. But, as we have seen, those 'same' things are unfolding differently when measured from the stationary system. In particular, moving clocks run slow compared to stationary clocks, so that the time indicated by the moving clock upon arrival of the flash is less than 1/299 792 458 s. Therefore, in order to measure the same value for the speed of light in the moving experiment it is unavoidable that the distance traveled by the light flash has decreased. In other words, the moving meter stick has contracted. Since the length of *anything* can serve as a ruler, the length of *any* moving objects contracts.

The line of arguments just outlined is straightforward and the conclusion is indeed valid. However, in the context of relativity (and possibly elsewhere) we need to be circumspect when it comes to seemingly intuitive and common sense arguments. In the following paragraphs we will give a few more scenarios that help to illustrate what it means to measure the length between two points. Physics is about measuring things and measurement is about comparison—physics is the ultimate comparative science. This includes mundane properties such as extension in space or, in other words, length, area and volume. It seems nothing more needs to be said except what was already mentioned above: the distance between two points is defined as the number of times a reference object (aka the ruler) has to be positioned end-to-end to connect the two points. In the case of the two points marking the opposite extreme ends of an object, this procedure defines the length (or width or height) of an object[10]. That is all for stationary points. When the length of a moving object is to be

---

[10] The procedure, by the way, also clarifies how to convert between units. Suppose two points A and B are separated by a distance of exactly 1 m, i.e. a single one-meter ruler is needed to bridge the gap between A and B. If you use a yard stick instead, you will find that one is not quite enough. Approximately, you need an additional 1/10 of a yard stick, 0.093 61 to be precise. So the conversion factor is 1.093 61 yard per meter.

determined, we must add the stipulation that the locations of any two moving points have to be determined simultaneously. Fair enough, because otherwise the length of a moving object could be anything (including a negative number), depending on when the position of the front is measured relative to that of the rear. For example, the length of your sleek Ferrari Starburst Rocket (did you know you had one?) speeding at $100\,000$ km h^{-1} along a race track on the other side of the Moon can be ascertained by having detectors measure when the rocket blocks a light beam crossing the track. The sensor that turns off at time t for the first time detects the arrival of the front of the rocket. At that same time, the rear is adjacent to the detector that just stops being shadowed. The distance between these two detectors, measured in the standard way, is the length of the moving rocket. At the risk of being redundant: the length of any moving object requires not only a standard length measurement but also the determination of whether two events at two different points in space occur simultaneously. We know we are in trouble now.

Just to be on the safe side, you decide to measure yourself the length of the rocket while it is in motion (relative to the track, not relative to you). For that purpose, a straight bar has been mounted parallel to the ship between two rods snugly attached to the front and rear of the rocket. Moving about deftly in your appropriately tethered space suit, you fit exactly 100 one-meter sticks between the inner sides of the two rods. And I mean mathematically exactly 100. Your rulers are of the same kind as those used to determine the spacing of the detectors along the racetrack, and these detectors operate with high precision and are as closely and regularly spaced as needed—obviously, we are running another thought experiment. Back to your measurement using the ruler: the length of 100 m that you found for the stationary rocket is called the *proper length*, which is the length of an object in the reference frame in which it is at rest. What about the outcome of the measurement on the track? The two detectors sensing the front and back of the passing rocket are separated by a distance that is about 0.43 μm shy of the proper length. That is such a small change that it can be neglected for most practical purposes. Nonetheless, the difference is not zero and thus is conceptually significant. The moving rocket has shrunk!

We will now see—with more rigor than was provided in the beginning of this section—that length contraction is indeed a logical consequence of the principle of special relativity. To that end we shift our attention to the odd contraption shown in figure 4.5, a modified light clock equipped with a propulsion system to make it just as fast as your elegant racing rocket. The fact is, aerodynamic styling is just for looks in outer space. Clunky it may look, but with the simple geometry and straightforward operation of the clock we have a much better chance to get at the essence of the phenomenon of length contraction. It also hopefully reminds you of the Michelson–Morley interferometer—it should! Here is what happens. A clever mechanism, located at point A in the center of both segments, emits simultaneously two brief light pulses. At rest, the two perpendicular arms of the double clock are identical in all respects, in particular they are of the exact same length. The fact that the labels for the two segments of the clock are different, L and L', may be seen for now as a simple means to keep them apart in our following discussion. Although it will turn out that what is the same when standing still becomes different when moving. In any case, for the

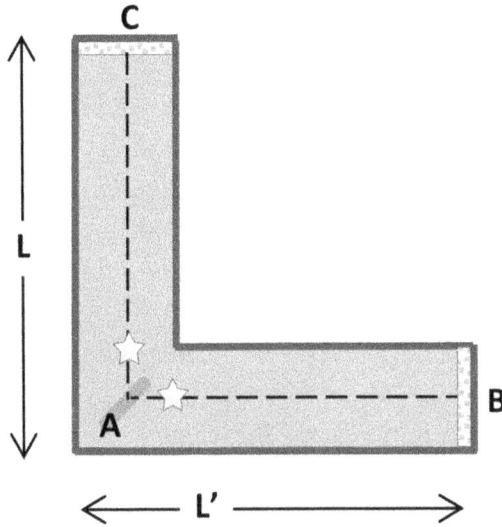

Figure 4.5. Double photon clock to detect simultaneity.

stationary clock L and L' are the same and thus, after reflection by their respective mirrors B and C, the two flashes return to A at exactly the same time. In other words, we have arranged for two mini-lightning bolts to strike simultaneously in one spot. Now let us see what happens when we set the L-shaped dual clock in motion in the direction AB and with constant speed v. Because the two events at A—the flash arriving from B and the flash arriving from C—are happening at a single location, the verdict of being simultaneous is absolute, rendered by any observer in any inertial reference frame. From our previous discussions we also know that (1) moving clocks run slow, (2) dimensions perpendicular to the motion must remain unchanged in length and (3) the speed of light is unchanged, as always. Therefore, in order to verify whether the return of the two light pulses remains simultaneous even if the clock moves, we must inspect the time it takes for the two pulses to return to their point of origin—when the clock moves. For the perpendicular segment ACA we have already done that for a single trip from A to C (see figure 4.3 and formula (4.1)). Since we use the complete round-trip here, all we need to do is to multiply the result we found earlier by two:

$$T' = \frac{2L}{\sqrt{c^2 - V^2}}.$$

The round-trip time for the parallel path ABA is made up of two contributions. First, it takes a time T_1 for the light pulse to reach mirror B, which itself travels a distance VT_1 before the light flash arrives. Then, on the return leg, the flash needs a time T_2 to get back to mirror A, which now approaches, thereby shortening the distance traveled by the flash by VT_2. Multiplication of the two times T_1 and T_2 by the speed of light yields the corresponding distance traveled by the light pulse. Therefore, we have the two identities

$$cT_1 = L' + VT_1 \quad \text{and} \quad cT_2 = L' - VT_2.$$

Solving for the two travel times yields:

$$T_1 = \frac{L'}{c - v} \quad \text{and} \quad T_2 = \frac{L'}{c + v}.$$

We know that the sum of T_1 and T_2 must be equal to one tick T' of the moving clock. Otherwise the required simultaneity of the arrivals of the two flashes is lost. Therefore, we must require that

$$T' = T_1 + T_2 = \frac{L'}{c - v} + \frac{L'}{c + v} = \frac{2cL'}{c^2 - v^2}.$$

When we compare the two results for the time period of the moving clock, there is only one way to make them the same. The length L' that is parallel to the motion must be different from the length L that is perpendicular. More specifically, we must require that

$$\frac{2cL'}{c^2 - v^2} = \frac{2L}{\sqrt{c^2 - V^2}} \Rightarrow L' = L\sqrt{1 - \frac{v^2}{c^2}} = \frac{L}{\gamma}.$$

As we have seen above, the factor γ is always larger than one. Therefore, the length L' is shorter than the stationary length L. In other words, the length contracts. And if the length of one object contracts, all objects must contract. The argument is completely analogous to the one we used to bridge the gap between the two statements 'one clock is running slow' and 'time dilates' in moving systems. In this section, we have encountered for the first time the deep connection between time and space revealed by special relativity. What is changed in the time domain (time dilation, slower clock) is compensated by an equivalent change in the spatial aspect (length contraction, shorter distance). We will see later that we can define a 'distance' between two events in such a way that its value is invariant for all observers in any inertial reference frame.

It is not always easy to put yourself into the shoes of someone else, but it is worthwhile to try. (I am talking strictly physics of inertial reference frames, though in my experience considering other viewpoints can be quite helpful in general.) The above discussion provides in some detail what that entails for the simple act of measuring length. Of course, changing the point of view becomes much more challenging with more demanding tasks. The frequent use of phrases like 'from the point of view' or 'as judged in reference frame XYZ' may seem to imply that special relativity is about observational differences. Not surprisingly, a frequent question in this context is 'but is time dilation (length contraction, lack of simultaneity, etc) real?' Without wanting—or for that matter being properly equipped—to enter a philosophical debate, I can outline my own pragmatic approach. To the student's question, I am inclined to answer 'Yes, these effects are real.' The assertion that 'an object has this specific length' or that 'there is this particular time interval between two events' only makes sense when a reference standard and a comparison procedure are defined. As long as those definitions are provided, the meaning of 'length' and 'duration' are unambiguous. To my mind, there is also nothing more 'behind' the meaning—at least as long as we exclude quantum physics (see chapter 3). A successful measurement yields the specification of the number of appropriate units of length, mass, time and so forth. This *is* what it means for an object to 'have' length or mass or for a process to 'have' duration—they 'have' the property only in as much as a comparison to a suitable reference can be made. I also understand that the term 'measurement' is not as anthropocentric as it may sound. Any physical process—the scattering between two objects, the absorption and emission of a photon by an atom—probes the system and thus constitutes a measurement. Furthermore, pertinent numbers can be recorded by instruments without human intervention and therefore measurement comparisons have nothing to do with perception or sensory illusion.

4.8 Why gravity must slow down clocks

Stories in physics often do not have a well-defined beginning (or end for that matter) and general relativity is no exception. Our briefest of surveys of this theory begins with the Apollo 15 Moon landing in 1971. One of the related NASA web pages [6] contains a video of the famous hammer and feather drop that mission Commander David Scott performed on the lunar surface. In the near-perfect vacuum there, heavy hammer and light feather fall at the same rate to the ground—just as Galileo had hypothesized more than 300 years before (see section 1.5). Galileo had based his prediction on careful observations using various objects sliding and rolling down smooth inclines—and very likely not, as urban myth would have it, dropping objects from the leaning tower of Pisa. Another falling object, an apple, supposedly sparked Newton's insight into the law of gravity that put Galileo's idea on a firm mathematical foundation. However, *why* free-falling objects of any kind fall at the same rate remained an open question until 1915, when Einstein found a most ingenious answer in his theory of general relativity.

When Einstein published his paper on special relativity in 1905, he knew right away that his new theory had only limited validity. While instantaneous acceleration of objects *can* be dealt with by special relativity, a systematic treatment of accelerated frames of reference is left out. It took Einstein another 10 years of intense work and the help of his friends Marcel Grossmann and Michele Besso, both gifted mathematicians, to surmount the roadblocks he had recognized from the start. Ultimately, though, the insight at the core of general relativity is not a feat of elegant mathematics (of which there is plenty), but an insight that bears witness to Einstein's genius. He saw what had been hidden in plain sight for more than a century.

In Newton's language, the acceleration of a body subject to a gravitational force is described in the following way:

$$ma = F \quad \text{with} \quad F = G\frac{mM}{R^2} = mg \quad \text{with} \quad g = G\frac{M}{R^2},$$

where a is the instantaneous acceleration of an object of mass m under the influence of a force F. In the specific case of gravity between two point-like objects of mass m

and M separated by a distance R, the force has the indicated form, with G being the gravitational force constant. Of course, real objects have spatial extension, they are not points. As Newton had already proved with his newly invented calculus, the above equation holds even in the case of finite sized spheres. Suppose that mass M is a large sphere of radius R—such as the Moon—and mass m is associated with a small, irregularly shaped object—such as a hammer or a feather—located at or near the surface of the sphere. Then the situation can be well approximated by the equivalent case of two mass points, located at the respective centers of gravity, subject to the same gravitational force as shown above. Provided that the small object moves only short distances—compared to the center-to-center distance between the two objects—up or down, then the force remains essentially constant. For example, the radius of the Moon is about 1700 km, so that a drop of about 1.6 m for a hammer and feather implies[11] a change of the gravitational force of about one part in two million. This argument contains a tacit assumption. The mass of the falling object on the left-hand side of Newton's second law is what should properly be called the inertial mass of the object, its resistance to change of velocity. The mass that enters the gravitational force law has *per se* nothing to do with acceleration because the force can express itself in static ways, for example in the stretching of a spring to which the object is attached. In that context, mass should be termed gravitational mass. Here mass plays the same role that electric charge plays in Coulomb's force law. In order to bring out the difference between inertial and gravitational mass, indices i and g are introduced as follows

$$m_i a = m_g g \text{ or } a = \left(m_g/m_i\right)g.$$

In other words, the hammer and feather fall at the same rate when their inertial and gravitational mass ratio is the same. Several experiments and observations—much more precise than the Apollo 15 stunt—have demonstrated the fact that this mass ratio is for all practical purposes equal to 1 for all substances that have been tested. Quite a remarkable outcome, considering that inertia is intrinsically something completely different than gravity. For those scientists who thought about the issue, it was a coincidence, a remarkable one maybe, but still just a quirk of nature. Almost everyone else did not even think about it. In Einstein's mind the fact that the mass ratio is unity revealed a deep connection between inertia and gravity. It is actually quite simple: inertial and gravitational mass have the same magnitude because they are one and the same.

The following thought experiment, a variation of Einstein's self-proclaimed happiest thought, illustrates the equivalency. Cut Galileo's windowless cabin (see section 4.3) out of the ship and move it with an attached rocket to the outer edges of the solar system. Out there the residual gravitational force of the Sun is negligible (and if not, move the cabin even further away). The cabin is pressurized and quite comfortable apart from the annoying habit of things to float freely about without the orderly

---

[11] Because the force depends quadratically on separation R, a change of one part in X in distance causes a change of two parts in X in the force. If you are versed in calculus you can see this relation immediately from the rate of change of force with changing distance $|dF/dR| = 2\ GmM/R^3 = 2F/R$ so that $|dF|/F = 2|dR|/R$.

influence of gravity pulling everything downward. That includes the captain who has just woken up. At about that moment, the rocket engine turns on again, accelerating the cabin along a straight line and with constant magnitude g. Mercifully, the acceleration is in the direction from cabin floor to ceiling so that everything—sextant, bottle of rum and captain—lands on the floor. By the way, the cabin is also perfectly soundproof and the rocket runs as smoothly as you could wish. That first feeling of floating in the air when the captain awoke is quickly forgotten and attributed to a bit of too much rum. As long as the acceleration continues, the captain has no reason to doubt that he is still on the ship in calm waters, just off the coast of Italy, his legs firmly on the cabin floor (okay, maybe still a bit shaky). The acceleration of the cabin creates an environment that is indistinguishable from gravity. There is no measurable difference between a uniformly accelerated frame of reference and one in a constant gravitational field. Gravity has the same effect as acceleration and thus Einstein pronounces them to be the same. Inertial and gravitational mass do not have the same value by accident or magic. The two quantities are identical; there is only *one* mass, thus explaining Galileo's dictum. As a consequence of the equivalence between gravity and acceleration, if you let a box free fall in gravity, i.e. let the box move downward at a constant rate of acceleration, then the inside of the box becomes an inertial reference frame. It is as if that downward acceleration cancels the upward acceleration that is equivalent to gravity. Indeed, several research facilities such as the Bremen Free Fall Tower or NASA's zero-gravity training plane take advantage of this fact.

While all this is quite remarkable, we can justifiably ask what it has to do with time and clocks. Soon after the captain has sobered up he picks up his studies. Today they involve checking the two clocks in the cabin—one bolted to the floor, and the other attached at height h to the ceiling right above the first clock. Comparing their rates, the captain cannot detect anything unusual. Not that he would expect any difference, since both clocks are at rest in the cabin. But what would an outside observer measure, who resides in an inertial reference frame and whose location is fixed at point P next to the path of the cabin? When the ceiling level passes by point P, the cabin has a speed v_1. Because of the cabin's acceleration, the floor passes by P at the later time t with somewhat larger speed v_2. Regardless of any detail, we can say with certainty that the two clocks are therefore *not* running at the same rate from the point of view of the inertial reference frame. Using the time dilation relation of special relativity, we can quantify this difference in clock speed the following way. Denote with t_1 and t_2 the rates of the floor and ceiling clocks, respectively, as measured from the inertial reference frame. Then , as long as the speeds involved are small compared to the speed of light:

$$\frac{t_1}{t_2} = \frac{\sqrt{1 - \left(\frac{v_2}{c}\right)^2}}{\sqrt{1 - \left(\frac{v_1}{c}\right)^2}} \approx \left(1 - \frac{1}{2}\left(\frac{v_2}{c}\right)^2\right)\left(1 + \frac{1}{2}\left(\frac{v_1}{c}\right)^2\right)$$

$$\approx 1 - \frac{v_2^2}{2c^2} + \frac{v_1^2}{2c^2} = 1 + \frac{v_1^2 - v_2^2}{2c^2}.$$

On the other hand, the distance h traveled between the two measurements is related to the acceleration g and the two speed values as $V_2^2 - V_1^2 = 2hg$, so that approximately

$$\frac{t_1}{t_2} = \frac{t + \Delta t}{t} = 1 + \frac{\Delta t}{t} \approx 1 + \frac{gh}{c^2}.$$

In other words, an observer in the inertial reference frame finds that the two clocks are running at different rates. This is not too disturbing since, as we just saw, the two clocks pass by the observer at different speeds. Now comes the fun part. According to Einstein's equivalence principle, an accelerated frame is indistinguishable from a frame in which the same gravitational acceleration prevails. Switching reference frames in the above example is straightforward: the two clocks are now at rest and separated in height by h. The erstwhile outside inertial reference frame is now free-falling downward with acceleration g. Because of the strict equivalence, the two clocks in the cabin must again show the same rate difference. As a consequence, after a time t they will show a difference Δt of

$$\Delta t \approx \frac{ght}{c^2}.$$

For a 3 m ceiling height, Earth's surface gravitational acceleration $g \approx 10$ m s^{-2} and an elapsed time of one second, the predicted effect is miniscule: $\Delta t \approx 3.3 \cdot 10^{-16}$ s. No wonder the captain did not find anything unusual. Of course, if you are more patient, the rate difference accumulates. You still need extremely precise clocks—such as the aluminum ion-based atomic clocks available to a group at NIST in Boulder, Colorado. In 2010, they were able to verify [7] the rate change predicted for height difference of as little as 30 cm[12].

As we saw in chapter 2, the period of any oscillator depends on certain parameters that characterize its physical make-up. Obviously, changing the value of any of these parameters changes the oscillation frequency. Conversely, the rate of a different type of clock that does not depend on the parameter in question is *not* affected. Gravity has already crossed our path in connection with the pendulum, but in that context it was not at a fundamental level. The dependence of the pendulum period on the strength of the local gravitational force is just an indication of how the pendulum works and has nothing to do with time itself. Now, though, this is different: general relativity predicts that gravity *does* affect the flow of time after all. In the presence of a stronger gravitational force *any* type of

---

[12] In truth though, Einstein's conjecture of time being slowed by gravity was demonstrated experimentally much earlier.

clock will run more slowly by the amount shown above[13]. And when all clocks run more slowly, time runs more slowly.

4.9 A pair of twins most famous

There is probably no textbook on special relativity that would not include at least a paragraph on the famous twin paradox. If you do not know the story, here is a made-up Twitter version (with a few characters to spare):

> Twins go on different journeys. Both claim their clock runs true. But when they reunite, one of the twins is older. #TwinsNoMore

Because of time dilation, changes in clock speed are to be expected. Because of the relativity principle, all clocks, including biological ones, will run more slowly. Since this observation depends solely on *relative* motion, and since the ether and absolute space are dead, both twins observe the same slowing in the other in a perfectly symmetrical way. However, when one of the twins returns after a looping trip to the point of departure, she finds—and this is the paradox—that her sibling who stayed at home has aged more. Truth be told, no humans, animals, or plants have ever been subject to any trip that would have induced noticeably differential aging. It is a different story for ultra-precise timepieces. Meet atomic cesium clocks A, B, C, D and E. You counted correctly, our siblings are actually quintuplets. In October 1971, atomic clocks A through D were taken on two commercial around-the-globe flights, one pair eastward and one pair westward. Sibling E stayed home at the Naval Observatory[14] to tell the reference time. At the end of their journey, the west-bound time of the traveling clocks lagged by 273 ± 7 ns behind the homebound clock. Considering the detailed groundspeed and altitude log of the various flight legs, the predicted time difference was 275 ± 21 ns. Similar (dis)agreement was found for the eastbound clocks [8] so that, within uncertainties, the experimental and theoretical results agree[15]. So you see that even if you did nothing but fly in airplanes all your life, it would not really help much in keeping you youthful. I bet it would backfire. Of course, what matters is the principle and there are instances when time dilation does matter. For example, cosmic rays—we will have to talk more about them in the following module on deep time—send showers of high-energy particles, mostly electrons and protons, into the upper layers of our atmosphere. When these fast particles collide with air molecules, sub-atomic particles called muons are

---

[13] Of course, the amount will differ and the relation will be more complicated when the gravitational acceleration cannot be assumed to be at least approximately constant. In particular in the vicinity of massive objects, near the event horizon of black holes, and similar situations, the relation above must be modified.

[14] This is an oversimplification. Comparison of the traveling clocks' time was made with mean US time as told by the observatory. But the observatory time is a weighted average from a fairly large number of atomic clocks, not just a single one. In other words, atomic clock E alone is a whole group of siblings.

[15] When the time difference that relativity theory predicts for the above experiment was calculated, not only the time dilation due to the speed of the traveling clocks had to be considered. Another contribution of about the same magnitude came from gravity's influence on time. In other words, it came from Einstein's second, his general theory of relativity.

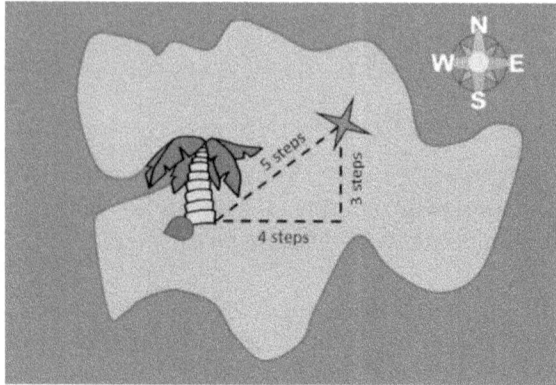

Figure 4.6. How to find the buried treasure.

produced, which continue the path of the incoming rays towards the ground. These secondary products are moving fast, but because they are short-lived, only a small number should survive the trip to Earth's surface. The number of muons actually detected at sea level is much larger than any non-relativistic estimate predicts and can only be explained by the motion-induced longevity of the muons: their inner clock runs more slowly and they live longer than their idle twins produced in the laboratory. From the point of view of the outdoor muons, the Earth rushes towards them just as the racetrack moves past the rocket. In their own reference frame, their lifetime is as short as it is in the Earth-bound laboratory. Nevertheless, they make it to sea level in abundance because the distance between the upper atmosphere and the ground has shrunk. This example illustrates how time dilation and length contraction emerge in various proportions, dependent on the specific reference frame, so that certain outcomes are independent of the reference frame, here for example the number of muons hitting detectors at a specific altitude. While time interval and spatial distance by themselves are variable, we will see next that a certain combination of the two is a unique and invariant measure of the four-dimensional, space–time interval between two events.

4.10 The invariant space–time interval

Everyone knows how to read a treasure map such as the one shown in figure 4.6. First, find the island, then the palm tree. Take four steps to the East, then three steps to the North. Start digging. Of course, if you know a little geometry and algebra, you can easily figure out the distance of the treasure from the palm tree by using the Pythagorean Theorem. Here, the horizontal distance comes out as five steps (square root of the sum of the squares of four and three). You can get the direction as well from the stated instructions, but that is of no concern here. If you want to know the three-dimensional, straight line distance ΔL between the starting point at the palm tree and the buried treasure, you have to include the depth at which the chest is buried. Using the labels Δx, Δy, Δz for the number of steps—or whatever length units you use—along the three spatial directions, Pythagoras tells us that ΔL can be obtained from the relation

$$\Delta L^2 = (\Delta x)^2 + (\Delta y)^2 + (\Delta z)^2$$

For example, a depth that is equivalent to 2 steps turns the 3D distance into the value of $\sqrt{29} \approx 5.4$ steps. Although not always the most practical and sometimes not even a possible route, ΔL is the shortest distance between start and finish. Lines connecting two points with the shortest possible distance are also called *geodesics* and the relation, such as the one above, that expresses the distance resulting from small displacements along the coordinates is called the associated *metric*.

Back to our adventure. When you eventually find the treasure trove, two events frame this success—taking the first step away from the palm tree and laying your hand on the gold coins. Of course, these two incidents cannot happen simultaneously. A finite, non-zero temporal interval, Δt, intervenes. We have the option to specify the two 'distances' in space and time separately or to combine them into a single quantity. In Newtonian physics, space is three-dimensional, and so it does make sense to combine length, width, and height into the overall spatial distance shown above. But space and time are distinctly different and no new physical insight is gained by combining the two into a single mathematical expression. Also, because spatial and temporal distances are measured in different units, either ΔL or Δt or both would have to be modified. One simple solution is the multiplication of time with a velocity so that the product has units of length. While there is no speed in non-relativistic physics that would lend itself as a universal multiplier, in special relativity the speed of light is a natural choice. Using that approach we might be tempted to write, as the space–time interval between two events, the sum $\Delta L^2 + (c\Delta t)^2$—a straight forward extension of the space-only case. This expression certainly seems reasonable as it grows with increasing spatial or temporal distance. The drawback is its variability from observer to observer, which follows from the transformation of length and time for moving observers discussed in sections 4.5 and 4.7. This may not sound very alarming since relativity is, well, about relativity. We can do better though! A simple change of sign[16], turns the above expression into a description that is invariant upon switching between inertial reference frames:

$$\Delta s^2 = \Delta L^2 - (c\Delta t)^2 = (\Delta x)^2 + (\Delta y)^2 + (\Delta z)^2 - (c\Delta t)^2$$

From this definition follows that Δs^2 can be zero or even negative. Thus, in order to avoid taking the square root of a negative number, the definition is given in terms of squares and Δs^2 is referred to as the relativistic space–time interval. But what does it mean if Δs^2 is zero or negative? Suppose two events occur at two separate points A and B in space with a time Δt between them so that, in your reference frame, the event at B happens after the one at A. Further suppose the event at B is something bad and you are a prime suspect to have done it. In order to make it specific, let us

---

[16] Notice that the definition used here subtracts the time-like interval from the space-like one. In the literature, the order is sometimes reversed.

suppose the crime, yes it is that bad, occurred at the train station in Boston at noon sharp. You need an alibi. Prior to the event in question you were in Auburn, in round numbers 70 km to the West. Since your car can travel as fast as but not faster than 140 km h^{-1}, you could get to Boston in half an hour (speed limit, traffic and safe driving be damned)—but not earlier. Luckily, you can prove that you left Auburn at 11:40 AM and so you could not have done it. Unless you borrowed your friend's Ferrari. Even that is not the end of the story since you might have operated a remote drone. In that case, the time between pushing a button in Auburn and causing a result in Boston might be as short as the time the electronic signal needs to travel 70 km. If the signal is carried by light in the open air, that time is not even the blink of an eye, a meager 233 μs. But that is it. Nothing that left Auburn with fewer than 233 μs to spare can cause *anything* at noon sharp in Boston! In special relativity, the light speed sets a natural and impenetrable boundary for two events to be causally linked to each other. And the condition that the space–time interval be zero defines this boundary. In our mini-crime story, the two events in question are 'leaving Auburn' and 'crime in Boston'. From the discussion of the drone case, it is clear that the term $c\Delta t$ is exactly equal to the physical distance ΔL between Auburn and Boston and therefore $\Delta s^2 = 0$ for this case. A null space–time interval is also called a light-like separation between two events. Should Δs^2 be positive or negative, the distance is declared to be space- or time-like, respectively. In the former case ($\Delta s^2 > 0$) the associated two events cannot be a cause-and-effect pair while the former condition ($\Delta s^2 < 0$) allows it. To be clear: the possibility that an event A caused event B does not mean it actually did, as our Auburn–Boston example shows. Here, the space time interval for *any* choice of car comes out to be negative as long as the departure time from Auburn is larger than the 223 μs needed for the light-like separation quoted above. You might have used the drone...

Hopefully, the significance of the space–time interval as defined has become clearer. What is not obvious right away, at least not to me, is the claim that Δs^2 is invariant, i.e. that its numerical value is the same for all observers (we might call them witnesses in this instance) regardless of their state of motion. For the general cases of space- and time-like separation, the proof is somewhat involved. But for light-like intervals, $\Delta s^2 = 0$, it is fairly easy to see its invariance. In a given reference frame, the zero condition implies that $c^2 = (\Delta L/\Delta t)^2$. Changing to a different reference frame, moving at speed v, the space and time separations transform to different values $\Delta L'$ and $\Delta t'$ according to the relations shown in sections 4.5 and 4.7 above. However, the speed of light is unaffected so that still $c^2 = (\Delta L'/\Delta t')^2$ preserving the null value of space time interval in the moving reference frame. That is the proof in this case, but the invariance does indeed hold in general.

In summary then, if certain things can or cannot cause other things in a given reference frame, then this possibility or impossibility of causation persists in *any* *other* reference frame. Specifically, the boundary between these two cases is universal. The invariance of the four-dimensional space–time interval is analogous to a square ABCD that is being observed from different two-dimensional coordinate systems which have point A as their common origin, but which are rotated relative to each other. In one system, the *x*-axis is parallel to the AB and CD sides of the square. One might say that these two sides have only 'x-character', while the BC and DA sides have 'y-character'. The length of the AB side, for example, is simply $L^2 = (x_B - x_A)^2 + (y_B - y_A)^2$ where $x_A = y_A = y_B = 0$ and $x_B = L$ are the corresponding point coordinates. Now let us rotate our frame so that the new x-axis, call it x', is parallel to the AC diagonal of the square. Of course, nothing has changed as far as the square is concerned, but from the viewpoint of the new frame, the squared distance between points A and B is given by $(x_B' - x_A')^2 + (y_B' - y_A')^2$ where now $x_A' = y_A' = 0$ and $x_B' = L/2^{1/2}$ and $y_B' = -L/2^{1/2}$. In that sense, side AB side has now both x- and y-flavor, although, as it should be, the length AB is still the same value L. In this spirit, it is often said that the space–time interval between two events is invariant, but that its admixture of a 'space-like' and 'time-like' component varies with the motional state of the observer. It is weird to think that the geometric distance between Auburn and Boston and the temporal interval between two events there is relative and subject to change. On the other hand, we have the consolation that the corresponding space–time interval is rock solid.

References

[1] Jenkins M 2015 How a remote peak in Myanmar nearly broke an elite team of climbers *National Geographic Magazine* (September 2015)

[2] Maxwell J C 1873 *A Treatise on Electricity and Magnetism (Part IV)* (Oxford: Clarendon Press)

[3] Michelson A A and Morley E 1887 On the relative motion of the Earth and the luminiferous ether *Am. J. Sci* **34** 333–45

[4] Pedersen K M 2000 Water-filled telescopes and the pre-history of Fresnel's ether dragging *Archive Hist. Exact Sci.* **54** 499–564
(for a pictorial explanation see: https://upload.wikimedia.org/wikipedia/commons/8/81/ Stellar_aberration_versus_the_dragged_aether.gif)

[5] Epstein L C 1993 *Relativity Theory Visualized* (Covington: Insight Press)

[6] http://nssdc.gsfc.nasa.gov/planetary/lunar/apollo_15_feather_drop.html

[7] Chou *et al* 2010 Optical clocks and relativity *Science* **329** 1630

[8] Hafele J C and Keating R E 1972 Around-the-world atomic clocks: observed relativistic time gains *Science* **177** 168–70

Chapter 5

Deep time or getting old

'How old are you? Really? You look much younger. Are you sure?' Well, how indeed do we know how old we—and things—are? The philosopher in us wants to add: 'How do we know anything?' Granted, this is a good point, but the physicist simply replies 'I do not know and my hunch is I never will. But I know how to measure and how to compare empirical data with each other and with concepts and theories.' So, in the spirit of this book, let us again see what we can learn by doing just that—measuring, comparing. Coming back to the original question, the answer about your age is rather straightforward if you happen to have a birth certificate. Presumably it will indicate the year, month and day of your birth, maybe even the time of day. Subtraction of the current date from the date of birth yields your age. How is this looking up of records a measurement, though? I claim it is and that it is perfectly equivalent to the way the time interval between two events is determined in the context of relativity: all we need is a set of synchronized clocks, one at the place of your birth and one here and now. The date entered in your birth certificate was in agreement with a valid calendar and possibly a local clock. As mentioned in chapter 1, calendars may be understood as the counting part of the celestial clock that ticks off days. Therefore, to the extent that calendars are accurately kept and read, the time between two events is easily measured to an accuracy and precision of about a day. When Julius Caesar waded with his troops across the Rubicon, a shallow river in Northern Italy, he crossed a much more important political line, changing history in the process. Caesar was fully aware of the consequences and famously uttered the words 'Alea iacta est'. That all took place on January 10 according to the Republican, pre-Julian calendar then in use and in the year 49 BCE by our current Gregorian calendar. Since conversion from the Republican calendar is uncertain, let's assume that Caesar had already introduced his calendar reform. Then while not really important, we can calculate the number of days that have passed since then

doi:10.1088/978-1-6817-4096-6ch5

and any particular day thereafter, say 10 January 2018—a number[1] that is independent of any specific calendar choice. Because, at one time or another, changes were made in the way we keep calendars in unison with the flow of the seasons, it can be a challenge to pinpoint the occurrence of events that lie father back. Recorded history entails different time spans for different cultures, but eventually we run out of calendars, at least those that are created and kept by humans. Thus, the idea of counting the days between two events seems to run out of steam when one of the events lies in the distant past. When no clock is present, no time measurement can be made. While this statement is true by definition, not all is lost because we have yet to verify that indeed no clock was present. As we will see in this chapter, clocks can come in unfamiliar forms that rely not only on periodic but also non-periodic, irreversible processes. In the latter case, time-telling is based on two distinct aspects, namely on quantifying *accumulated change* occurring at a *rate* that must be known. First, though, we will see how far back you can go with the help of birth records. It turns out, all the way to the beginning—or so James Ussher believed.

5.1 Earth's age—a cautionary tale

We have a clear notion on when the Earth coalesced out of the debris of our Sun's predecessor star: four and a half billion years ago is consistent with plentiful and varied evidence pertaining to our planet's age. Our fair planet has existed for quite a while—much longer than anyone had thought until about a hundred years ago. In the 1920s, a new dating method, based on the measurement of radioactive decay products, gained traction, and started to persuade many that the age of the Earth numbers in the low billions of years. Not the hundreds of millions that geologists were willing to allow as necessary for the slow geological processes to unfold. Much more than the tens of millions that some physicists had estimated around the same time. And, certainly, a lot more than James Ussher [1] had calculated way back in the early 17th century.

Usher was born 4 January 1581. When he was 26 years old, he became Professor of Theological Controversies at Trinity College in Dublin, Ireland. Quite uncontroversial was the concept that the Earth and everything in, on and above it had been created and thus had a beginning—exactly when that had happened was not clear, though. How could you even attempt to find out how old the Earth is? Of course, if records existed that chronicle all times from the creation to the present, all one must do is add the numbers. In the Christian belief system, such a document exists. The bible contains a narrative that spans the entire time from the creation of Earth and the cosmos to the time of Christ's death. Setting his faith in the literal veracity of these stories in the Bible (and using one specific version of the various ones in existence), Ussher set out to add the lifespans of all generations since Adam and Eve to his own time. Where the bible has gaps or is vague, he consulted other historical documents. The result of his meticulous study is an astonishingly precise date of the

---

[1] 754 594 if you must know. I used the Julian Date Converter provided on-line by the US Naval Observatory. Quite amazing that some words are remembered for so long.

creation of the Earth: 22 October 4004 BCE (when using the Gregorian calendar backwards). Ussher was neither the first nor the last person trying to pinpoint the age of the Earth, but somehow his attempt stuck and remains one of the better-known variations on this theme.

For quite a while, the historical method must have appeared quite reasonable to those few people who even thought about the age of the world. There simply did not seem to be anything else one could do. However, clues already existed in plain sight that hinted to a possibly much older age of the Earth. In plain sight, as in a landscape near you or as in fossilized marine animals found in the mountains. James Hutton, an 18th-century geologist of many skills and callings, was involved with the building of canals and on his site visits he had a chance first to observe and then to really see what many others had only looked at—that in many places the ground we stand on has easily discernible layers, characterized by different types of rocks, color, composition, texture and so on. He was the first to challenge Ussher's pronouncement. In his three-volume book *Theory of the Earth* [2], Hutton applied the idea that physical and chemical processes acting today have also been acting in the past and are responsible for any and all geological change, an approach now commonly called uniformitarianism (as in *physical processes unfold uniformly the same way—then and now*). In 1795, when he published his first volume, this was a revolutionary thought—perhaps it is not a coincidence that around this time other radical ideas swept through Europe. In any case, Hutton's opaque writing style severely limited the immediate influence of his work. It took another 40 years before a lawyer turned geologist, Charles Lyell, brought these ideas to the attention of a much wider audience with his compendium *Principles of Geology* [3] (a lucid modern explanation of Lyell's propositions is given in [4]). Based on extensive and careful observations, Lyell concluded that the Earth must be ancient for the doctrine of uniformity to have a chance to explain what we can find on Earth today. If Hadrian's Wall, built by the Romans, had not disintegrated into dust in 1500 years, Earth must be older than the 6000+ years found by Usher—a lot older! During the 19th century, estimates of the age of Earth made in the new scientific spirit proliferated but were also all over the place. However, numbers in the range of millions of years were quite common. Charles Darwin, for example, in the first edition of his *Origin of Species* [5] estimated that the erosion of a certain geological feature in south-east England had to have taken place over at least 300 million years. Such large time spans are so vast that it is difficult, maybe impossible, to develop a feel for them. In any case, 300 million years did not sit well with many people—not just followers of Usher. Several physicists had joined the geologists (and naturalists like Darwin) in the new sport of estimating Earth's age. One of the most notable contestants was William Thomson (born 1829, since 1892 Lord Kelvin) [6]. He approached the issue by assuming that the Earth was initially a hot molten blob and slowly cooled to its present state. And that takes some time.

Glass blowing is an amazing art and requires impressive skills. Start with simple ingredients, heat them up to just the right temperature, inject air, and rotate and move the melt constantly—if you do it right the most beautiful, intricate, delicate objects emerge upon cooling. If you do not—and there are many ways to get it wrong—you might end up with a mess on the floor. *That*, of course, happens because of Earth's gravity. So, what would happen to a sphere of molten glass on the International Space Station or better yet in outer space, far away from the gravitational influence of any other body? First, depending on whether the hot, molten object rotates or not, it will acquire the shape of a flattened or perfect sphere. Once the energy input that caused the heating stops, the sphere starts to cool and eventually solidifies. Size, composition and initial temperature, they all influence the time it will take for the sphere to solidify completely. For a while a solid crust will coexist with a liquid interior whose diameter shrinks as time goes by. While the outside surface cools primarily by radiation—a relatively simple and well-understood mechanism—the cooling of the interior is more complex and involves thermal conduction as well as convection. With the best available data for the thermodynamic properties of rocks, Thomson first came up with a wide-ranging estimate of 20–400 million years for an Earth-sized, completely molten sphere to cool to the present state with a hot, liquid core lying about 2900 km below the surface. After several refinements, the by then Lord Kelvin settled on a value of about 20 million years for the age of the Earth.

Lyell and Kelvin were not alone in trying to get at the age of Earth and at the end of the 19th century different age estimates existed that were based on established facts and reasonable arguments. All ranged from a few to hundreds of millions of years. Compared to the numbers derived by physicists, those obtained by the geologists tended to be larger, with Lyell's estimate at the high end. But which number was the right one? Since scientists could not agree among themselves, maybe Usher was right after all? The resolution of this discrepancy turned out to be in favor of the geologists, but it came from a new discovery in physics. At around the time Lord Kelvin announced his 'short' lifespan of 20 million years, radioactivity was discovered, which—in the context of our story here—accomplished two things. First, it helped to explain why the age of the Earth is underestimated in the thermodynamic calculations. Radioactive decay in Earth's interior creates heat, which counteracts the straightforward cooling processes, and so it takes the Earth much longer to get where it is now. In other words, when radioactive heating is included in the thermodynamic models, the age of the Earth increases[2]. But what turned out to be even more important in our present context was the fact that some scientists very quickly realized that changes in the composition of rocks caused by radioactive decay can be used to *measure* the age of rocks. After a few false starts and dead ends, the qualitative idea turned into a reliable method producing quantitative results. At the end of this development stands the current best estimate of about four and a half billion years—15 times larger than even Lyell's appraisal. So, what is this mysterious radioactivity all about? All in good time. First, we shall take a detour.

---

[2] Truth be told, an improper account of heat convection in Earth's interior has also been pinpointed as a reason for the erroneous results of the physicists' aging model.

5.2 Built on sand—the hourglass as an analogue to radioactive dating

All of the clocks we encountered in the first three chapters of this book have one aspect in common—they are based on repetitive processes. What limits the performance of any clock, in more or less subtle ways, is the fact that they are interacting with the world outside themselves, which introduces irreversible, stochastic changes in the clock cycles. This deviation from perfect periodicity is usually seen as unavoidable and as a manifestation of the unidirectional march of time. We all know—in our minds and bodies—that time is an arrow. Elegant and simple, the hourglass is the embodiment of the notion of fleeting time (figure 5.1). Once inverted, the clock mimics and marks the disappearance of time by a smooth, steady flow of finely grained sand. A narrow bottleneck that separates the two sections of the hourglass slows the pace at which the upper reservoir empties out. Many other methods have been used to delineate irreversible, *linear* time. By burning incense packed into strings, sticks, or carved grooves you can not only tell how much time has elapsed, you can smell it. Substitute water for sand, add a few technical details, and you have turned the hourglass into a clepsydra, a water clock.

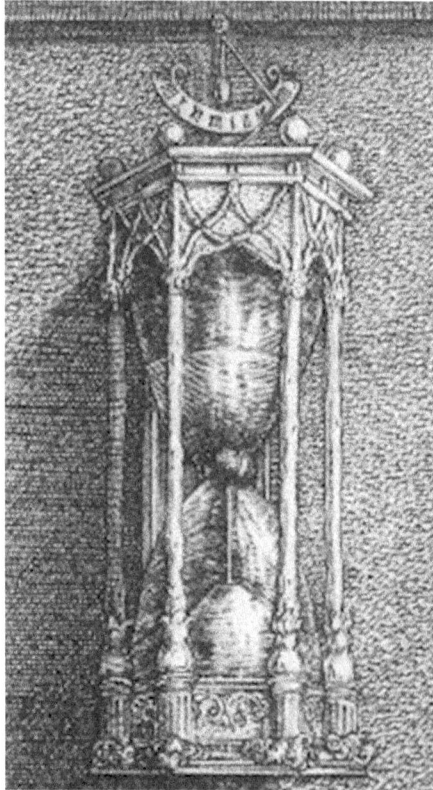

Figure 5.1. Hourglass in Dürer's *Melencolia I*.

Historically, it happened exactly in reverse order: the water clock is much older than the hourglass. Apparently, the first references to the latter appear only after 1300 CE, while water clocks were used in ancient Greece and Rome. Even older is the archeological evidence of clepsydras in China, culminating in the stunning clock designed by Su Song, which also serves to mimic astronomical processes. As varied as these beautiful timepieces are, here we only focus on the one feature that unifies them, their conceptual essence. For starters, even if a smooth flow of time were to exist, linear clocks cannot capture it since any medium we happen to use in our clock is intrinsically discrete. The passage of a single grain through the neck of an hourglass or the addition of one more water molecule to the clepsydra's measuring container constitutes the smallest unit of time these devices can discern. On the other end, the longest temporal interval we can determine depends on how slow the substance flows and how large an amount the receiving container can accommodate. That is, of course, where the *Guinness Book of Records* comes in. Appropriately housed in a museum dedicated to sand, in the Nima district in Japan, the largest hourglass in the world needs as much as one full year to completely drain its fill of sand [7]. Someone counted the grains in this misnomer of an hourglass and came up with about 630 billion. Thus, on average, about 20 000 grains flow through the neck in one second. Assuming you can discern a single grain, the smallest unit of time you could tell with this clock is therefore of the order of $1/20\,000$ s $= 50\ \mu$s. The qualifier 'on average' is important because the flow rate of sand in an hourglass is not necessarily uniform. For a not too viscous liquid, the flow rate through a small hole is governed by the pressure, which in turn is proportional to the weight of the fluid column above the hole. Granular material has a more complex flow behavior [8] than liquids, and local pressure is only one determining factor for the flow rate. The shape of the granules and their size relative to the bottleneck also play a role. Still, with care—and tinkering—hourglasses can be constructed to work at least approximately as smoothly as assumed above.

But what does all that have to do with the advertised topic of telling time by radioactive decay? Not much, on the surface. However, just as we can convert an hourglass into a clepsydra, so can we transform it into a new type of clock that *does* have a lot in common with radioactive decay. Here is how we accomplish this transmutation. First, imagine that the sand grains change color when they arrive in the receiving part of the hourglass—to be specific, assume that they change from green to red. Other than looking pretty, this color trick does not change anything about how the clock functions. However, you can tell whether a grain has moved from one state (being in the upper container, signified by the color green) into another (being in the lower container, signified by the color red) just by color discrimination. This feature allows us to simplify the clock in two steps. First, we dispense with the neck and place the entire volume of sand into a single container. That is, of course, a fairly dumb thing to do if your aim is to build a clock based on something changing. Good point. Let us therefore introduce a second modification: even though the grains stay where they are, we still let them turn from green to red. For now, we do not inquire *why* the sand grains behave in this chameleon-like fashion (we will get back to this question). All we need to know is that (1) the grains change color randomly and irreversibly, and (2) they do so at a rate that is equal to the rate at which grains pass through the neck between the two compartments of the regular hourglass. For example, 630 billion grains flickering randomly from green to red at a rate of 20 000 grains per second but lying perfectly still are equivalent to their flowing cousins in the year glass at Nima. Assume that at one point in time all of the grains are green. A year later, they are all red. At any moment in-between, if you can somehow establish the ratio of red versus green grains, you have found out how much time has passed since the start of the color show.

Replace 'grains of sand' by 'carbon atoms' and 'color' by 'mass' and you have a crude model for a widely used dating method based on the radioactive decay of unstable nuclei. While I need to explain a few things before all this makes more sense, the essence of any type of radioactive dating is just that: provided you know how quickly nucleus X decays into nucleus Y, a measurement of the present amount of Y relative to that of X settles—at least in principle—the age of the object since the mass transmutation began.

It will be useful to provide a mathematical description for the way an hourglass tells time—whether by flow or color change. Let us denote by N_0 the total number of grains, by t the time that has elapsed after either the turning of the hourglass or the color change has commenced (so that $t = 0$ indicates the moment of time just before the first occurrence of change) and by r the rate of change (whether flowing into the lower container or changing color). Then the number of grains that have changed from their initial state (being in the upper container or being green) to their final state (being in the lower container or being red), $N(t)$, is given by the product of rate and elapsed time, i.e. $N(t) = r \cdot t$. As it should be, the number in the final state is zero at time zero. As is, the formula allows the number of grains in the final state to become larger than the total number available, which is, of course, impossible. We can easily fix this flaw in our mathematical model by limiting the time for which the formula is valid. At some time, call it τ or the *lifetime* of the sand stored in the upper reservoir, the number of grains in the final state will be equal to the total number, i.e. $N(\tau) = r \cdot \tau = N_0$. From this latter relation it immediately follows that the lifetime of the hourglass is numerically identical to the ratio of total number of grains and flow rate, i.e. $\tau = N_0/r$. By the way, this expression for the time to empty out the initial state completely is exactly what I mentioned above: the maximum time an hourglass can tell time is proportional to the amount of sand and inversely proportional to the flow rate.

5.3 Survival graphs and aging

While the analogy between the hourglass and radioactive decay is qualitatively okay, there are important quantitative differences that can be illustrated with the help of figure 5.2, which depicts the demise of three cohorts initially containing 100 000 individuals each—one of them comprising grains in an hourglass, with a second comprising radioactive nuclei and a third comprising human males (mostly for dramatic effect and to show that things can be different). The horizontal axis ticks off years and the vertical axis tallies the number of surviving individuals. It is

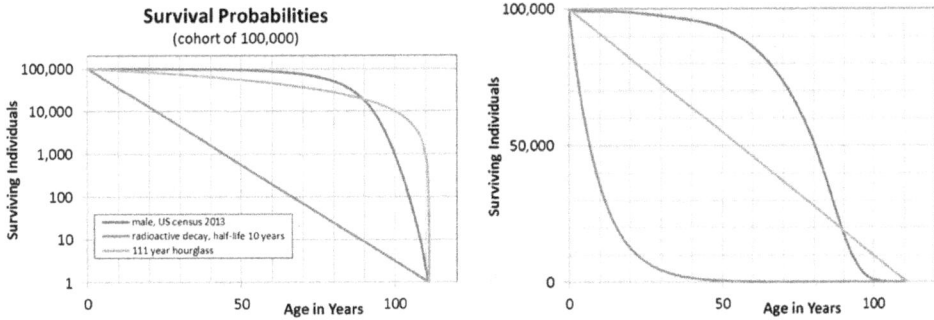

Figure 5.2. Survival curves for three different cohorts of initially 100 000 individuals (see text and legend for details).

the choice of the third group that dictates the range of the time axis: according to the 2013 census of the US Social Security Administration, after 111 years only one out of the 100 000 individuals will statistically be alive. The numbers of survivors for all other years are taken from the pertinent table and are summarized by the blue curve in the figure. If we take an hourglass that contains 100 000 grains and is shaped so that the last grain falls into the lower container after 111 years, the drainage of the initial amount of sand follows the green curve[3]. On the other hand, a certain radioactive substance with a half-life of about 6.7 years[4] will trace the red curve. Figure 5.2 comes in two versions showing the same information plotted in different ways. While both axes of the right panel mark the variation of the corresponding quantity in a linear fashion, the vertical axis of the left panel, i.e. the number of surviving individuals, is logarithmic. As can be seen, the grains of sand in an hourglass diminish in a straight line when both axes are linear as, of course, it should be. The fact that the sample of radioactive nuclei depletes in the shape of a straight line in the left panel indicates exponential decay. Obviously, the change of the population of humans is different.

One way to discuss these differences in population change is to concentrate on the *rate* of change. We have already seen that sand grains fall at a constant rate—at least that is the approximation we use here. If one of our color-changing sand piles is to mimic the decay associated with radioactivity it must behave quite differently and in the following manner. Regardless of how many original-color grains remain, half of them change to the final color during a time span that is always fixed in length. Thus, in terms of absolute numbers, more radioactive nuclei convert at the beginning than near the end, which is exactly what the red curve in figure 5.2 illustrates. The rate is *not* constant. One can show that an exponential decay fits the required bill.

---

[3] This will be a highly peculiar *hour*glass. To satisfy the given constraint, we need to engineer a device that releases one grain every nine and a half hours or so. No big deal if we can apply Einstein's *Gedankenexperiment* approach. For example, making the neck sufficiently narrow, long and/or sticky should do the trick.

[4] We will see shortly what exactly is meant by characteristic lifetime. Also, the radioactive nucleus coming closest in lifetime might be cobalt-60 (5.3 years). So, here as well, we may have to apply unrestrained engineering.

Empirically we may write—using the same notation as above for the hourglass case—the number of nuclei remaining in the initial state at time t as $N(t) = N_0 \cdot e^{-t/\tau}$. Therefore, after the characteristic time τ, the number of survivors is given by $N(\tau) = N_0 e^{-1} = N_0/e$ so that the corresponding fraction is $N(\tau)/N_0 = 1/e \approx 0.368$ or (very) roughly one-third. After a time of 2τ, the fraction of remaining individuals is $1/2e$ and so forth. The contemporary US male and—with only small difference—female population ages in the opposite way: with relatively low infant mortality (it still could be better), the death rate is much smaller at the beginning of the overall time span than near the end. Lest you think that the strikingly different exponential decay is characteristic for inanimate 'aging', consider that the survival graphs (that is what the curves in figure 5.2 are called) of songbirds and lizards exhibit a striking similarity to that of radioactive nuclei. Of course, that formal resemblance is only superficial and in any case accidental. As is the case with humans, the mortality rates of animals depend strongly on environmental conditions. In contrast, no known process (such as exerting pressure, exposing to high temperature or harsh chemical substances, etc) will affect the characteristic lifetime of a given isotope[5]. At any given time, the probability for a radioactive nucleus to 'die' is the same no matter how much time has passed since it was 'born' and no matter what the environmental conditions.

5.4 The neutron and other unstable characters

It is time to investigate in a little more detail what this radioactivity decay is all about—for example, how it manages to change the mass of atoms or to transform their chemical character. It sounds like alchemy's dream of turning lead into gold—which it very nearly is. As we saw in module 3, atomic clocks take advantage of the dynamics of the electron cloud surrounding the nucleus in each atom. Radioactive dating relies on changes of the inner core of the atom—the nucleus with its protons and neutrons. In 1898, Pierre and Marie Curie [9] coined the word 'radioactivity' (to be a stickler, it was *radioactivité*), referring to the spontaneous emission by certain substances of ... well what exactly these substances were emitting was not immediately clear and the catch-all label 'radiation' had to suffice. Two years earlier, their fellow physicist Henri Becquerel had discovered this phenomenon when uranium salt blackened a photographic plate that was completely wrapped in paper, thus preventing exposure to ordinary light. Further complicating the issue, only a few years before that, in 1895, Wilhelm Röntgen had found a type of radiation—x-rays—that had similar effects. It took a while and many separate investigations by several scientists before a clear image emerged of the new physics manifesting itself in these phenomena. Radioactivity is the spontaneous transformation of a nucleus, accomplished by the emission of particles and/or electromagnetic radiation and in accordance with energy and momentum conservation. In the end, radioactivity is yet another incarnation of the generic trend of systems to move towards their lowest

---

[5] Which is not entirely correct—bombardment with high-energy particles can stimulate decay. However, the collision energy has to be extreme, so that the nuclear structure is affected, a feat that of now can only be accomplished in accelerator facilities.

internal energy state. Eventually, a kicked pendulum will come to its resting point (see chapter 2), excited atoms give off photons and revert to their ground state (see chapter 3), exothermic chemical reactions run their course towards equilibrium (maybe prompted by a little tickle—like shaking that nitroglycerin) and certain nuclei give off energetic particles or photons to drop into more stable configurations. As alluded to above, tickling is neither necessary nor effective in the case of radioactivity. Radioactivity that proceeds by particle emission is called either α- and β-decay, depending on whether it involves the expulsion of a helium nucleus or an electron, respectively. Rather unsurprisingly, the remaining type of radioactivity is labeled γ-decay. The net effect of the first two types of radioactivity is to reduce the number of neutrons in the nucleus (by one in β-decay and by two in α-decay) while simultaneously increasing the number of protons by one (β-decay) or lowering it by two (α-decay).

In chapter 3, we discussed the various types of elementary particles that are the building blocks of all of chemistry and by extension biology: electrons, protons and neutrons. Isolated electrons and single protons are completely stable[6] as far as we currently know—in other words, they live forever. Not so the neutron. When trapped in suitably arranged magnetic fields[7], it only survives on average for about 880 s, not even 15 min, before spontaneously turning into a proton and emitting a fast electron in the process[8]. How can we picture this spontaneous disintegration? A crude image might be as follows. The neutron—as with any other particle that decays—has an internal structure; it is not a featureless point-like object. Furthermore, the structure is dynamic in the sense that constant fluctuations occur in the spatial arrangement of its composites (quarks and gluons in the case of the neutron). Suppose that, on average, a time span T lies between two fluctuations and that one in N configurations of the internal parts leads to decay. Usually, this goes along with a lowering of the intrinsic energy of the decaying object, which necessitates the emission of some other particle (electron, alpha particle, photon, you name it). In any case, an average time of NT will elapse before the break-up occurs. Depending on sample time T and the rarity of the unstable configuration, the decay constant can be expected to assume just about any value imaginable. Think about the simple mechanical clock introduced in chapter 4: a ball bouncing elastically between two walls. If there is nothing more, the system is stable. Now imagine that the poor ball is trapped in a prison cell with a barred window in one

---

[6] A proton *can* disappear but only in collisions with other particles. For example, when a proton meets its evil twin, the antiproton, they annihilate each other. The same occurs in the encounter of electrons and positrons or for that matter between any particle and its antiparticle.

[7] Although electrically uncharged, the neutron possesses a magnetic moment, just like a tiny bar magnet.

[8] Neutron decay is not quite as simple as that. More detailed experiments revealed a conundrum: the sum of the energy and momentum of the electron and proton is different from that of the neutron. Something is amiss. Eventually, the missing quantities were attributed to an additional particle, the neutrino—or, more precisely, the electron anti-neutrino. Neutrinos interact only extremely weakly with ordinary matter and easily escaped detection in the early experiments.

wall. The space between the bars is smaller than the diameter of the ball. So still no escape. But if the bars are not completely rigid and randomly swing a bit sideways, thereby changing the space between them, then there is a chance that this fluctuation will allow the ball to escape from the cell. Equivalently, random changes of the ball diameter or a magic and spontaneous splitting of the ball into two smaller ones can accomplish the same result. Again, depending on how often the fluctuations occur that allow escape and how frequently the ball visits the barred window, a sentence to life without parole may end much sooner. The key ingredient in the above story is the word *random*. Thus, the moment for decay (aka escape) may occur right away or only after a long time. Of course, the image sketched above should not be taken too seriously. Quantum mechanics, for example, does not allow for such simple time-varying paths and it is not clear how to attach in general meaning to terms like *configuration* or *time between fluctuations*. Still, the image might help a little. Let us return to the neutron.

The short lifespan of the neutron seems to bode ill for the stability of any chemical element other than hydrogen, with its lone proton at the center. On the other hand, both deuterium and helium, which contain at least one neutron in their nucleus, have been around for most of the age of the Universe. Since even Mr Ussher would agree that this is at least a few thousand years, we have a seeming contradiction here. The paradox is resolved by observing that the neutron in *any* atomic nucleus has the close company of protons and, except in a few cases such as deuterium, other neutrons. The proximity of like-minded nucleons stabilizes the neutron. Until it does not. From all the elements, lead is currently acknowledged as the heaviest stable element. In other words: if the crowd (of protons and/or neutrons) grows too large, the stabilizing influence is diluted and the nucleus becomes unstable, just as for a lone neutron. Even for elements below this threshold for instability, radioactive decay occurs if the neutron number deviates from the optimum. For example, the nucleus of the carbon atom contains six protons. In this case, two *stable* isotopes exist: carbon-12 and carbon-13 with six and seven neutrons, respectively, with the former isotope being more abundant by a factor of 10. There are also unstable isotopes, of which the longest-lived is carbon-14, with a half-life of $\sim$5730 years—which is long compared to human life expectancy, of about the same duration as recorded history, and short compared to any cosmic time scale. All other carbon isotopes decay much more quickly and none of them occur naturally. Carbon-14 does, which is peculiar.

Because carbon-14 has a lifetime that is so utterly short compared to the age of the Earth, this radioactive isotope has no right to be present at all if it was created at Earth's birth, let alone before that time. The depletion factor is the inverse exponential of the ratio of Earth's age (4.5 billion years) to the carbon-14 lifetime (5730 years) or approximately $1/e^{789\,000}$, which is zero for all practical purposes. Nevertheless, together with its vastly more abundant two stable siblings, carbon-14 exists in small but readily measurable traces, bound in atmospheric carbon dioxide. To be precise, almost one in a trillion (10^{12}) carbon atoms is of the elusive persuasion. How radioactive carbon gets into the atmosphere is a fascinating story having to do with cosmic rays that constantly bombard the upper layer of Earth's atmosphere, turning nitrogen-14 isotope into carbon-14 nuclei[9]. From there, the newly produced carbon reacts with oxygen to form CO_2 and diffuses to ground level, where plants and animals (us included) incorporate it into their bodies. Once the organism dies, the fraction of radioactive carbon dwindles exponentially at the above-quoted rate. Suppose that stable carbon is red and radioactive carbon is initially yellow, turning transparent upon decay. Then, the entire carbon reservoir initially has a slight orange hue that slowly vanishes after the death of the organism. Measuring the degree of 'orangeness' reveals how much time has passed since the sample of organic material stopped the uptake of carbon dioxide. Of course, after a sufficiently long time the remaining carbon-14 concentration is so low that further decay does not produce a measurable change. Therefore, from a practical point of view, objects older than a few half-lives, say between 20 000 and 50 000 years old, cannot be dated reliably with this method. As is the case with every type of measurement, the more precise and accurate your instruments, the more carefully you use them, and the more insightfully you evaluate the circumstances under which the sample incorporated the carbon, the more reliable your age assessment will be[10]. What is measured in practice is not the actual number of carbon-14 atoms, but the number of radioactive decay events per unit time. As we have seen above, these two quantities are proportional to each other in the case of exponential change. Since it is much easier and can be done with much higher precision, electrons resulting from the β-decay of carbon-14 nuclei are detected for a given amount of total carbon and during a predetermined length of time. With the numbers given above, it is easy to show that for one gram of carbon in equilibrium with the atmosphere, almost 1000 decay events occur per hour (~ 936 h^{-1}). For a sample that is 39 000 years old (or more accurately that died this many years ago) and that contains one gram of carbon, on average only one decay per hour occurs—we are approaching the above stated useful limit of the carbon-14 dating method.

5.5 Radioactive dating of rocks

In a recent brochure, the United States Environmental Protection Agency warned that you cannot see, smell, or taste it, but that it may be bad for your health. The culprit is radon, a radioactive, inert gas that poses a health threat not just in the USA but worldwide. If a certain amount of this gas is captured in a closed container (say the basement of your house), it will disappear with a half-life of less than four days, provided no new gas enters the container during this time. The problem is that the latter is hard to do in the case of houses: if at some time radon managed to enter somewhere, the chances are that there is a practically inexhaustible supply of it in the

---

[9] Cosmic rays consist mostly of energetic electrons and protons. While the particles originate outside the solar system, their flux through Earth's upper atmosphere is modulated somewhat by solar magnetic activity. A detailed understanding of the carbon-14 production rate is important and still subject to current research, mostly concerned with isolated events of extremely high or low rates. Also, the above-ground atomic bomb tests performed from 1955 to 1980 by various countries have injected a measurable spike of carbon-14 into the atmosphere.

[10] Apparently, recently killed freshwater mussels appear to be much older because their carbon-14 concentration is lower because of depleted material such as limestone.

ground below and all you can do is to minimize infiltration or improve ventilation. In 2000, the General Assembly of the United Nations received a report on atomic radiation that included data on radon levels reported as countrywide averages. Levels of this naturally occurring gas differ by a factor of 10 or so between countries, with Sweden and Finland showing some of the highest values. This variability hints at a geological dependence of radon abundance that does indeed exist. Radon was one of the first radioactive elements to be discovered, identified very quickly as a product in the decay of uranium. It may sound a bit alarming that uranium is widely distributed in the Earth's crust as a component of many minerals. However, unless it is enriched in localized ore deposits, it is only a trace component, buried safely in the ground. Depending on the type of rock, uranium concentrations vary—that is how Sweden and Finland ended up with their high radon reserves. Overall, uranium and several other radioactive isotopes are responsible for the Earth-warming energy release that I alluded to in the opening paragraph of this chapter. How long has this been going on for, you ask? For an astonishingly long time. That is why uranium radioactivity is a good geochronometer for determining the age of ancient rocks with good precision, as we will see next.

First, though, we should address the question of what it means to speak of the age of a rock. Diamonds are forever and while our planet Earth, the third rock from the Sun, was not made of that precious stuff, it was made of stuff nonetheless. So, are rocks ageless? Remember our intrepid geologists Hutton and Lyell? They and their posterior colleagues figured out the recipes for how to make just about any rock you can find. The corresponding cookbook is quite a compendium, but there are three basic recipes for igneous, metamorphic and sedimentary rocks, referring to the most recent process by which the contemporary piece of solid came into existence. Given how old the Earth is, plenty of opportunities existed for rock formation and, therefore, the age of stones and minerals varies widely. Nevertheless, the oldest rocks we find provide a lower bound of Earth's age. At the time of writing this book, the birthday of the oldest known mineral was 4.3 billion years ago, about the same age as that of lunar rocks brought back from the Moon by the Apollo missions. At about 4.5 billion years, certain meteorites are slightly older. Since they seemed to have coalesced during the earliest phase of the solar system, at around the same time as the Earth solidified, their age is taken to be that of the Earth[11].

Just like carbon-14, any radioactive substance will eventually deplete to a negligible amount. How long this will take depends of course on the specific decay time. Certain nuclides maintain their activity much longer than radioactive carbon, some for many millions of years. Among the elements that have isotopes with very long lifetimes are uranium and potassium. In round numbers, ^{235}U, the most common uranium isotope, has a half-life of 4.5 billion (yes, billion) years. The

---

[11] Chondritic meteorites, to be specific. The oldest sample was recovered from a meteorite that came down in Mexico in 1969 and has been dated at 4.567 Ma old. In the end, the 'birth' of Earth—or any other planet for that matter—does not occur in a single moment, but is a process. When that process precisely begins is a question of definition. For Earth, a further complication arises in the form of the cataclysmic collision with another protoplanet that resulted in the formation of our Moon.

corresponding number for potassium ^{40}K is 1.25 billion years. Thorium, with a whopping 14.5 billion-year lifetime, is another element that is important in the context of deep-time radioactive dating. For many unstable isotopes, the immediate product is another radioactive isotope, which in turn may decay to yet another unstable nucleus. However, after a finite number of steps, the sequence will terminate with a stable product. For the uranium isotope ^{238}U, this termination point is lead, ^{207}Pb[12] and for potassium ^{40}K it is argon ^{40}Ar. Transformation to the stable element is often complex because intermediate products can decay in more than one way and because many steps may be needed before stability is reached. For example, the decay of uranium ^{238}U into lead involves no less than 17 unstable intermediate nuclides. Some of these in-between products live a fleeting existence—once born, they disappear in matters of milliseconds or less. Others, however, linger for hundreds and even thousands of years. Given this complexity, it might seem that it would be difficult to calculate, at any given time, the composition of an initially pure sample of uranium. However, over the huge time spans that uranium-containing minerals have been around, each and every unstable intermediate product has vanished—even those with the longest decay constants. Thus, the underlying idea of extracting the age of a rock from uranium activity is not only simple in principle but can be used in praxis. For each uranium nucleus lost, one lead nucleus eventually appears, and so the ratio of lead-to-uranium in the sample increases steadily with time. If rocks contain uranium and are at least about 10 million years old, determination of the lead-to-uranium ratio r reveals the age t of the rock according to the so-called *age equation*, which follows directly from the exponential decay law that holds for radioactive decay [10]:

$$t = \tau_{1/2} \cdot \frac{\ln(1 + r)}{\ln(2)},$$

where $\tau_{1/2}$ is the half-life of the uranium isotope in question. Note that at the half-life, i.e. at $t = \tau_{1/2}$, the ratio is $r = 1$ by definition and thus $\ln(1 + 1)/\ln(2) = 1$, as it should be. Similarly, at $t = 0$, no lead is yet present and $r = 0$, which is consistent with $\ln(1 + r) = \ln(1) = 0$. While it is conceptually straightforward, the implementation of the method is technologically challenging, mostly because of the low amount of uranium present in typical rocks. Since potassium is more abundant, this decay chain is usually easier to use and gives more precise results. Ideally, more than one radioactive dating method is applied to one sample so that consistency can be checked. In this fashion, the oldest rocks found on Earth have been dated.

So far, so good. But as the old saying goes: the devil is in the detail. While the basic idea of radioactive dating is sound and as simple as the age equation suggests, its application to real-world samples is not without pitfalls—to say the least. And the older the rock, the higher the stakes, and the more controversial it gets. To be clear: the controversy is not about the approximate age of roughly 4–4.5 billion years, but issues such as how precisely we can pinpoint this age or which site has the bragging

---

[12] Here and elsewhere, I use standard notation for the chemical elements. Lead, for example, is denoted by Pb from the Latin *plumbum* (which is just what it sounds like when a chunk drops to the floor).

right to be host of the oldest rock on Earth. Take, for example, the mineral zircon, a common silicate-type gem stone that is found in many places in the world. The Jack Hills in Western Australia can boast to have yielded the longest surviving rock fragments ever found [11]. Zircon contains traces of the transition metal zirconium (the name of the element is derived from the mineral) as well as uranium. Therefore, the U–Pb decay can be used to date the age of zircon, which is what scientists of course did. The current best number for the formation of a few slivers of zircon is 4.4 billion years, with an impressive confidence limit of about plus/minus eight million years. It might sound misguided to call an uncertainty of eight million years a precise result, but this range constitutes a relative ambiguity of just $\pm$ 0.2%. One of the lunar rock samples collected by the Apollo missions is slightly older than the Jack Hills find. *In toto*, the evidence of careful radioactive dating of a variety of rocks is consistent with an age of the Earth of about 4.5 billion years. I think it is quite amazing that we learned to read Earth's age with such certainty. Yet this is not the end of the story. There is one more chapter to tell of an even grander accomplishment.

References

[1] Ussher J 1658 *The Annals of the World* (London: Crook and Bedell)
Irish Philosophy—James Ussher www.irishphilosophy.com/2016/01/04/james-ussher/

[2] Hutton J 1795 *Theory of the earth: With Proofs and Illustrations* vol 1 (Library of Alexandria)

[3] Lyell C 1853 *Principles of Geology: or the Modern Changes of the Earth and its Inhabitants Considered as Illustrative of Geology* (John Murray)

[4] Gould S J 1987 *Time's Arrow, Time's Cycle: Myth and Metaphor in the Discovery of Geological Time* vol 2 (Cambridge, MA: Harvard University Press)

[5] Darwin C 1968 *On the Origin of Species by Means of Natural Selection* (first publ. 1859) (London: Murray Google Scholar)

[6] Burchfield J D 1990 *Lord Kelvin and the Age of the Earth* (Chicago: University of Chicago Press)
Lamb E 2013 Lord Kelvin and the age of the Earth *Sci. Am.* (June 26) http://blogs. scientificamerican.com/roots-of-unity/lord-kelvin-age-of-the-eart/

[7] www.guinnessworldrecords.com/world-records/largest-hourglass

[8] Aguirre M A, Grande J G, Calvo A, Pugnaloni L A and Géminard J C 2010 Pressure independence of granular flow through an aperture *Phys. Rev. Lett.* **104** 238 002
www.technologyreview.com/s/418993/the-mystery-of-sand-flow-through-an-hourglass/http:// arxiv.org/abs/1005.2884

[9] Fröman N 1996 *Marie and Pierre Curie and the Discovery of Polonium and Radium* Nobelprize.org Nobel Media AB 2014 www.nobelprize.org/nobel_prizes/themes/physics/ curie/

[10] *Radiometric Time Scale*, US Geological Survey https://pubs.usgs.gov/gip/geotime/radio-metric.html

[11] Oskin B 2017 Confirmed: oldest fragment of early Earth is 4.4 billion years old *Live Science* Feb 23
www.livescience.com/43584-earth-oldest-rock-jack-hills-zircon.html

Chapter 6

From beginning to end

Although an age of 4.5 billion years for the Earth is certainly respectable, our quest to find and date the earliest events in nature does not end here. There are things we can observe and measure today that take us farther back into the past by almost another 10 billion years. Chemical elements and their isotopes were significant players in the previous appearance and disappearance act. Now, their abundance still plays a supporting role, but the main cast no longer resides on Earth but rather dwells in the heavens. Our story of telling time began with the stars (and planets and their satellites). It is a pleasing symmetry that it will also end in the realm of the stars—if it ends at all.

6.1 The fixed stars in their crystal sphere revisited

Can you tell if a star comes closer or moves further away by looking at it? That question may sound silly on two levels. First, we know that the *fixed* stars (we are not talking about planets here) are called that for a reason—they are supposed to twinkle for eternity in the position they occupy. Furthermore, even with the best telescopes currently available stars remain point-like light sources and therefore no change in size due to relative motion can be expected. When you spot a T-rex in the rear-view mirror of your car, you *can* tell whether you are in trouble by observing if the giant lizard's image in the mirror becomes larger or smaller. That just will not work with stars. Stars really do appear to be fixed in position. But are they? How could you find out? In principle, you could discover stellar motion the same way a police officer finds out whether you are speeding. Send out a sequence of constantly spaced radar pings in the direction of the star and then listen to the return pings. If the time intervals between the received pulses are longer than those between the emitted pulses, the star is moving away from us, while if the intervals shrink the star is approaching, and if they are unchanged there is no radial motion. However, if you want to apply this approach to objects outside the solar system, you

doi:10.1088/978-1-6817-4096-6ch6

must be patient—very patient. Even for Proxima Centauri, the closest star, the radar signal will take almost seven years to return. And for our companion galaxy Andromeda the round trip lasts a whopping five million years plus change. But more importantly, the fraction of returned pings will be prohibitively small. We need another method.

Quite a while ago, in 1842, Christian Doppler had an idea. At the time Doppler published his thoughts [1], observational astronomy had discovered many new and fascinating objects in the sky, including bound systems consisting of pairs of stars[1]. While the color of individual stars is often quite subjective, the proximity of two stars in the field of view of the telescope made it quite clear that the light emitted by these binary stars had different colors. By the mid-19th century, the electromagnetic wave nature of light was firmly established—we encountered this theory in the context of atomic clocks in chapter 3, and some pertinent details may be found there. One of the important results of this theory is the association of color with the wavelength or frequency of light. Doppler showed that these attributes of a wave depend on the state of relative motion between the emitter (the star) and the detector (the eye of the astronomer behind the telescope). Specifically, receding motion causes the crest-to-crest distances (aka wavelength) to increase linearly with relative speed, while approaching motion bunches the crests up and thus leads to a linear decrease of wavelength with relative speed. In the color spectrum, the wavelength of blue light is approximately half that of red light. Therefore, Doppler speculated that stars moving towards us have a bluish tint (have a blue shifted color spectrum) while a reddish color (redshifted spectrum) results from motion away from us. While Doppler correctly pointed out the effect—which now bears his name—that the relative speed between emitter and observer has on the wavelength of the electro-magnetic radiation, his use of that fact to explain the color of binary stars was soon dismissed and is indeed misplaced[2]. Thus, around the mid-19th century, we still did not know much more about stellar motion than the ancient Greeks and a static, eternal cosmos continued to be consistent with the evidence. After several more decades, though, an unexpected observation shattered this state of affairs. As we will see shortly, the redshift of extremely distant objects in the sky can be fruitfully interpreted as being due to the Doppler effect after all, and thus as a sign of a dynamic cosmos. The change of our view of the Universe that came with this insight turned out to be as revolutionary as the one Copernicus had brought about half a millennium earlier.

---

[1] It is amusing, that in most modern physics textbooks the Doppler effect is introduced as an acoustic phenomenon—exemplified by the infamous ambulance siren (higher pitch when it approaches, lower pitch when it drives away). Clearly, the title of Doppler's paper indicates that acoustics was not his primary topic. However, in the paper, he did extend his new theory right away to sound.

[2] Suppose that the relative speed between a source and a detector is given by v. Then the relative wavelength (frequency) change registered by the detector is given by $\Delta\lambda/\lambda = \Delta f/f = v/c$, where $\lambda(f)$ is the wavelength (frequency) of the light emitted by the same but stationary source and c is the speed of light. Individual stars can only be observed within our galaxy where stellar speeds relative to us are insufficient to account for observed colors which are determined primarily by their surface temperature.

By the end of the 19th century, several influential practitioners had started to transform astronomy from a craft that was mostly—if not exclusively—concerned with measuring and cataloguing the position of stars, planets and a smattering of other celestial objects into a discipline with much greater ambitions. Two new goals had been established, namely to gain a deeper understanding of the properties of stars and to map out the grand structure of the Universe. Tools and techniques from physics were adopted for that purpose. Most prominently, breakthroughs in optics and photography [2] allowed astronomers to collect starlight with unprecedented efficiency and to decompose it into its constituent colors. What Herschel, Fraunhofer and others had already done with sunlight could now be accomplished with starlight. After Kirchhoff and Bunsen had discovered that the dark lines in the spectrum of our Sun are the telltale signs of chemical elements in the outer layers of the Sun, it quickly became clear that starlight contains the signature of the same elements—hydrogen chiefly among them—that we find here on Earth. As we have already seen in chapter 3, each element has its own spectroscopic *fingerprint*, i.e. a unique list of wavelengths for which emission can be found. We also know, since the establishment of quantum physics, that these wavelength markers have their origin in the internal structure of the elemental atoms. Hydrogen atoms in any star—no matter how far away—are the same as those in my (or your) laboratory. That, at least is, our modern view of the world: one set of rules for all. Therefore, the expectation is that the presence of hydrogen gas should be identifiable by a strong emission or absorption line in the red portion of the visible spectrum, more precisely at a wavelength of 656.3 nm, the famous Balmer α-line. Likewise, sodium goes along with two lines at 589.0 and 589.6 nm, respectively. If the emission lines are strong, as is the case, for example, with calcium, the corresponding element does not even have to be particularly abundant.

It did not take very long before the spectra of thousands of stars had been recorded and classified. With improved sensitivity and better methodologies, it was only a small step for astronomers to look for spectroscopic signatures in non-stellar objects as well. As early as 1912, Visto Slipher, who worked at the Lowell Observatory near Flagstaff, Arizona, noticed that the emission lines visible in the light from Andromeda, our companion galaxy, were shifted by a noticeable amount. He also realized that the Doppler effect can explain this shift very easily. While the observed shifts in wavelength were unambiguous, they were also small—about one part in a thousand. Since the relative wavelength shift is equal to the ratio of the emitter–detector speed to the speed of light (see footnote 2 of the present chapter), the corresponding relative speed between our galaxy and the Andromeda nebula is immediately found to be about 1/1000 the speed of light or 300 km s^{-1}. This speed is 10 times larger than the speed at which Earth hurtles around the Sun. Much too fast a pace for an entire galaxy, at least for the taste of many contemporary astronomers, who dismissed Slipher's interpretation right off the bat. But as is the tendency with inconvenient facts in science and elsewhere, they did and do not go away. Within a few years, Slipher and other observers had measured the Doppler shifts of about two-dozen spiral nebulae. There was good evidence that members of this class of objects have even larger speeds relative to us than the Andromeda nebula. Because the early and thus limited survey of galaxies showed a few—Andromeda included—moving towards us and some moving away, Slipher had developed the idea that our Milky Way is embedded in a stream of nebulae, approaching us on one side and receding on the other. However, with a database that was growing in quality and sample size, it became apparent that, in all directions of space, the vast majority of spiral nebulae move away from us. Panel (a) of figure 6.1 gives an idea of the pertinent data that was available in the late twenties: the apparent 'escape' velocities of extra-galactic objects, measured in units of km/s, is plotted with their distance from us, in astronomical distance units of mega parsec, abbreviated as Mpc, equivalent to almost 3.3 million light years or $3.1 \cdot 10^{19}$ km. As can be seen from the dashed line in the figure, measured values can loosely be fitted by a linear relation between velocity and distance. Why would that happen?

6.2 The cosmic egg

Georges Lemaître was the first person to link the observed flight of the galaxies with the idea of an expanding universe or, more precisely, expanding space in which objects move faster the further away they are. Imagine a yeast dough (aka space) with raisins (aka galaxies) dispersed throughout the dough. The shriveled grapes have no choice but to follow the rising dough and, for any given raisin not too close to the edge, the others are moving away in all directions. In a 1927 paper [3], Lemaître explained how Einstein's general relativity equations allow for solutions that describe such an expansion of space itself[3]. In this model, the redshift of the galaxies comes about because all galaxies move away from each other, being taken for the ride of their lives by space itself. A one-dimensional toy model can help us to see what is supposedly going on. Imagine a rubber string with nine pearls (yes, you guessed it: string=space, pearl = galaxy) evenly spaced at intervals of 10 cm and no pearls at the ends so that the whole string is 100 cm long. For the analogy to work, we must assume the pearls to be tiny, say with a diameter much smaller than 0.1 cm—not really what you want in a necklace, but that is not what we are after here. Hold the ends of the string and slowly, say over the span of one minute, stretch it at a constant rate so that, at the end of this not very aerobic exercise, the string is 110 cm long and therefore the distance between neighboring pearls has increased to 11 cm. Now comes the interesting question. During the expansion, what are the relative speeds between the pearls? The answer depends on the distance between the pearls under consideration. For two neighboring baubles, the distance has changed by 1 cm in one minute so that their relative speed is obviously 1 cm per minute. However, in that same time span, the distance between the two next-nearest neighbor pearls (i.e. with one other pearl in-between) has changed by twice the distance, namely by 2 cm, and thus their relative speed is 2 cm per minute. For every additional pearl spacing, an additional length change of 1 cm must be added, resulting in a relative speed that increases linearly with distance. Voilà—Lemaître's conjecture! Well, almost.

---

[3] The Russian physicist and mathematician Alexander Friedman had already shown in 1922 that for an isotropic and homogeneous universe, Einstein's field equation of gravitation has solutions that are characteristic for a universe that either expands or contracts [4]. Lemaître was unaware of that work.

To make the analogy even closer, we must stretch our imagination a bit more by letting the initial length of the rubber band and anything on it shrivel to zero[4].

For both the rubber band analogy and the real thing, the expanding universe, there is no local motion of the embedded objects—pearls or galaxies—*through* space. For example, you can fix the pearls to the string using glue. Because of the insertion of space, otherwise stationary objects appear to move away from each other at a speed that increases in linear proportion to their distance. Nevertheless, the evidence is indistinguishable from that which moving objects would produce and thus an analysis based on the Doppler effect is valid.

At that time Einstein favored a stationary universe—as did almost anyone else—and to that end he had introduced into his field equations what one might call a fudge factor. It is thus perhaps no surprise that when the master met Lemaître at a conference, he supposedly said something akin to 'nothing wrong with your mathematics, but your physics stinks' (this is *not* a direct quote). In scientific debates, you should be prepared for harsh critique. But, of course, you are also entitled to the expectation that specific reasons be cited why your arguments offend someone's olfactory organ. So, what is wrong with Lemaître's idea of an expanding universe as the cause for the observed redshifts? Nothing, really, and in the end his idea carried the day. However, and in fairness to Einstein, we must say that the database for Lemaître's model was quite limited. Furthermore, while he predicted on theoretical grounds a linear relation between speed and distance of galaxies and even came up with a numerical value for the proportionality constant, Lemaître did not query very carefully the observational data for such a relation. That feat was accomplished, two years later, by Edwin Hubble. As already mentioned above, figure 6.1 illustrates the observational data that were initially available. Astronomers

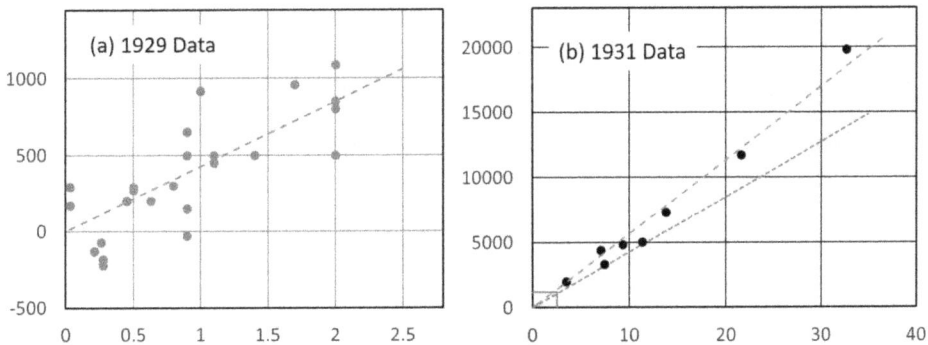

Figure 6.1. Velocity (km/s) of galaxies on the vertical axes as a function of distance (Mpc) along the horizontal axes: 1929 database shown in panel (a) and outlined as red rectangle in panel (b) which shows results available in 1931. The slope of the straight line fit is the Hubble constant.

---

[4] How close to zero? If the rubber band contracts at a constant rate from a finite length, mathematically it takes a finite amount of time to reach zero length. Thus, it is possible to start with zero. However, physically the idea of a singularity is fraught with problems as no current theory speaks to the physics under conditions of infinite energy density, which is what you get if you pack the existing matter, whether rubber band or cosmos, into a single point.

usually depict the same information slightly differently by using the *observed* Doppler shift on the *x*-axis rather than the *inferred* velocity (see footnote 2 in the present chapter). In any case, the data suggests the linear relationship that is included in figure 6.1 as the dashed straight line best fitting the data. The slope of the linear regression line is now known as the Hubble constant, with units of km s^{-1}/ Mpc^{-1}—a unit of measurement only astronomers can love. For good reason, though: divide the observed escape velocity by the Hubble constant H_0 and you have the distance of the galaxy. Figure 6.1 contains two panels with subsequently larger data range with Hubble's initial analysis based on the data shown in part (a). Already a few years later, two dozen or so additional galaxies at 10 times the maximum distance had been found and analyzed. Panel (b) of figure 6.1 illustrates the situation and compares the new data to the early prediction, the red dashed line extrapolated from the 2.5 Mpc range of the 1929 data set to the one available in 1931. While the numerical value of the slope changed somewhat, the original idea of a linear relation held up. As a matter of fact, today, almost 90 years and a few hundred additional studies later, the linear relationship still holds for galaxies that are more than a thousandfold more distant than the furthest objects included in the early investigations. Hubble and company were clearly on to something! However, the value of the Hubble constant has evolved to a significantly smaller value than the one extracted from the 1929 or 1931 data sets. Figure 6.2 recounts this part of the story by plotting the various values of H_0 as a function of the year of the corresponding study. Distance measurements are difficult for objects that are very far away. As it turned out, early methods and calibration points had to be revised which explains the initial large H_0 values. But since the mid 1970s, the average value of the Hubble constant converges steadily to an ever more precisely determined value of 70 ± 10 (km s^{-1}) Mpc^{-1} or 70 ± 10 (km/Mpc) s^{-1}. The latter form of the units of H_0 is more compatible with the notion of space creation: it implies that every second, along any length of 1 Mpc of the emptiness between galaxies, 70 km worth of brand-new space appears. That is equivalent to about 2.8 cm (~1 inch) appearing

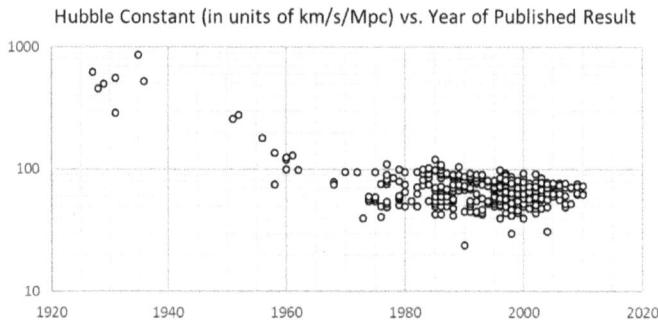

Figure 6.2. Hubble constant versus year in which study was published. Data taken from compilation by J Huchra, at the Harvard-Smihtsonian Center for Astrophysics as part of the NASA/HST Key Project on the Extragalactic Distance Scale (www.cfa.harvard.edu/~dfabricant/huchra/hubble/).

every year in a distance equal to that between Earth and Moon[5]. From a pragmatic point of view, the distinction between the two explanations of the redshift is not that important, since the creation of space can be viewed as the cause for an *effective* motion of the galaxies associated with a commensurate *effective* Doppler effect.

There is another important aspect of the observations: redshift depends solely on distance and not on direction, i.e. it is isotropic. In a moment, we will see why this is relevant. *If* the Doppler effect causes the observed redshift, whether effective or not, the observation implies that in any given direction galaxies move away from us with a speed that increases linearly with their distance from us—just as pearls on a rubber band that is being stretched.

But is the Doppler effect the only explanation? And, very importantly, how do we know how far galaxies are from us? Answers to these questions are not obvious and early doubters of the expanding universe theory had a reasonable point. As of today, however, these and other objections have been resolved in favor of Lemaître and Hubble. In the next paragraphs, we will explore a few of the pertinent lines of argument and corresponding observations.

For now, let us assume that galaxies are indeed hurtling away in such a way that their velocity v increases linearly with their distance D from us, i.e. $v = H_0 D$. In your mind, flip the switch that reverses the flow of time (it is the one labeled by the green and red arrows). As a result, all galaxies are approaching us—the dough contracts and moves the raisins towards each other. Since an object at a distance D approaches with speed v, it takes a time D/v. But since speed is given by the Hubble law, $H_0 D$, the time to join the fun, i.e. to arrive at our location, is the same for all galaxies, given by the inverse of the Hubble constant $1/H_0$. Since the rate of expansion is the same in all directions, all galaxies and all matter within our observable horizon merge and land on top of each other in one massive pile-up. Simultaneously, space itself has shrunk so that this gigantic collision takes place in the smallest of volume with a density of matter and energy that tends to infinity, a condition that mathematicians call a singularity. At the instance of perfect union, reverse the time direction switch once again to replay the movie in its original direction. Then you see that from this extreme point our present universe evolves— no further ingredient is needed, nothing is added except, of course, space. Lemaître referred to the initial entity from which everything follows as the *primordial atom* or the *cosmic egg*. Maybe he chose those names, in particular the latter, because he was also a Roman Catholic priest and had in mind the notion of creation. In any case, these names did not catch on. Instead astronomer Fred Hoyle's mocking character- ization of the new ideas as the *Big Bang* turned out to be the term that stuck[6].

---

[5] The presence of mass, even relatively small ones associated with planets and their satellites, slows the expansion rate dramatically so that space expansion is locally insignificant or even nonexistent inside systems with inhomogeneous, anisotropic mass distributions such as galaxies or planetary systems such as our solar system. See [5] and [6].

[6] Many languages have adopted the English term Big Bang, simply adding the proper article—such as *le Big Bang* in French (really). In German, the term *Urknall* is being used, which translates loosely as *Ancient Bang*.

From the above discussion it follows that the inverse of the Hubble constant is our first estimate of the age of the Universe. Using the currently accepted value of 70 km s^{-1} Mpc^{-1} yields a value of about 14 billion years since $1/H_0 \approx (3.1 \cdot 10^{19}$ km$)/(70$ km/s$) \approx 4.4 \cdot 10^{17}$ s ≈ 14 billion years. We will revisit this number when we discuss some finer points, but it turns out to be a reasonable value. When we compare the age of the Universe to that of Earth, about 4.5 billion years, we realize that our planet has been around for a good fraction of the life of the cosmos. On the other hand, there is a vast stretch of time in which Earth and the entire solar system did not even exist. A peculiar thought that one might find both unsettling and comforting.

6.3 Supporting evidence 1—ancient light

Arno Penzias' and Robert Woodrow Wilson's primary job was to test new telecommunication techniques[7] using a satellite called Echo II. Their bosses at Bell Laboratories had agreed that they could use the detection equipment afterward for astronomical observations. What eventually became modern-day satellite communications had, in 1964, quite humble beginnings: the satellite was nothing but a balloon launched into a low orbit with a metalized skin to reflect the microwaves aimed at it from the ground. Two solar cell-powered beacons, broadcasting at around 136 MHz, were the only onboard instruments. Penzias and Wilson attempted to detect the reflection signal that should occur when they sent microwaves in the direction of Echo. This echo off Echo was expected to be faint, but they had at their disposal a state-of-the art microwave detector coupled to a large and sensitive horn antenna at Holmdel, New Jersey, especially constructed for the Echo project. As sensitive as the detector was, persistent noise hindered the search for the return signal. A pigeon nest with assorted by-products in the antenna was first suspected, but in the end failed to be responsible for the noise. After a thorough cleaning of the horn antenna, the irksome microwaves persisted, coming from everywhere they looked. Annoyance turned into delight when the two researchers realized that the microwave background had a much more exotic origin. They had received a long-distance communication from space itself. The existence of microwave radiation had been anticipated as the remnant of the Big Bang radiation, cooled by the intervening expansion to a temperature[8] that was predicted to be about 3 K.

The story of the cosmic microwave background goes roughly like this: after the very first moments of the Big Bang, the cosmos had cooled to the point that the basic constituents of matter—electrons, protons and neutrons—had already 'crystalized' into existence. However, the cosmic climate was still too hot for these elementary particles to further coalesce into neutral atoms. Photons interact strongly with

---

[7] According to Wikipedia (https://en.wikipedia.org/wiki/Project_Echo), the Echo satellite program also provided the astronomical reference points required to accurately locate Moscow. This improved accuracy was sought by the US military for the purpose of targeting intercontinental ballistic missiles. I suppose this possibility for both civilian and military use is a good reason why space programs run by adversarial governments are seen with suspicion. Been there, done that.

[8] The fact that radiation can have a temperature was discussed in chapter 3.

charged particles and thus, because of the constant scattering, light was unable to travel far. The dense primordial fog only lifted after the expanding space had cooled the space filling plasma to between about 10 000 and 3000 K, the temperature range at which protons and electrons can recombine into neutral atoms. From that moment on, about 380 000 years after the Big Bang, most free electrons were removed from the environment so that light interacted only weakly with matter and could roam freely through the Universe. Therefore, using the age of the Universe from the Hubble constant, the cosmic microwave photons that are being detected today[9] were born more than 13 billion years ago. Truly ancient light—although, as we saw in chapter 4, in the reference frame of photons time stands still and for them not a second has elapsed. Remarkable as that is, even more astounding was the observation that the frequency distribution of the cosmic microwaves has the form of a thermal radiation of 2.7 K in *any* direction in space, in other words the radiation is extremely smooth and isotropic. After several improvements in observational tools and techniques—ranging from measurements taken with the U2 spy plane, balloon flights over Antarctica and several satellites, including NASA's COBE and ESA's Planck missions—the verdict is clear: the deviation from the mean value of about 2.7 K is at most a few tens of micro kelvin or a few parts in 100 000. Figure 6.3 reproduces the most detailed survey map currently available from the Planck mission. It shows the full celestial sphere in a projection that is also used to render the surface of the Earth on a two-dimensional map: the top and bottom points are the north and south celestial poles, respectively, and the horizontal line follows the celestial equator. Color indicates the tiny temperature deviations below (blue) and above (red) the mean with, as mentioned above, the largest differences measuring only a few ten millionths of a kelvin—an impressive feat of precision measurement. The map tells two important stories. It reveals that the cosmic radiation has an

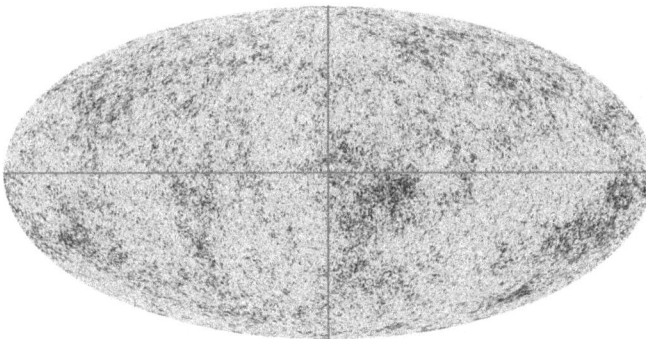

Figure 6.3. Tiny temperature fluctuations of the cosmic microwave background radiation, rendered by false colors in this figure, are indicative of corresponding matter density changes in the early universe. Source: ESA and the Planck Collaboration. See: http://www.esa.int/spaceinimages/Images/2013/03/Planck_CMB

---

[9] According to one source [7] cosmic microwave background photons cause about 1% of the noise in television sets that still happen to receive their signal through the air. So, if you want to stay in touch with the Universe, get rid of cable and dishes.

extraordinarily smooth and isotropic distribution, but then it adds that the distribution is not quite perfect after all. We will see next that the near-flawless isotropy is an indication of a brief phase of inflationary growth of the Universe, while the ripples are seen as harbingers of the spatial structure of the present-day universe at corresponding angular scales.

The map in figure 6.2 represents light from edge of the visible universe, light that was emitted soon after the Big Bang. Therefore, two opposite points of the horizon, for example two patches near the opposite celestial poles, are separated by a distance that is twice as large as light can travel during the entire age of the Universe. Since nothing travels faster than light, the two patches cannot have been in communication with each other. In other words, no known mechanism could have ensured that these two regions are in equilibrium with each other. Yet, as we have seen above, they are: their respective microwave radiation has the same temperature to within a ridiculously small variation. How can this be? In the early 1980s, several theorists—Alan Guth of MIT, Andrei Linde of Stanford and Paul Steinhardt of Princeton among them—proposed a mechanism that would account for the smoothness we see today. In the blink of an instance, starting at around 10^{-36} s after the Big Bang and lasting a whopping 1 to $10 \cdot 10^{-32}$ s, the rate of space expansion accelerated exponentially. Brief as this so-called *inflation* period was, it caused the scale parameter, the cosmic yardstick, to increase stupendously. Patches of the growing cosmos that were in 'contact' before inflation were out of reach of each other when it ended. Thus, the smoothness of the cosmic microwaves reflects the condition of the cosmos *before* inflation, while the small crinkles are the imprints of quantum fluctuations *during* or slightly after this epoch.

So, what about these faint ripples in the cosmic microwave pattern? What do they tell us, if they tell us anything at all? Are they not just noise? Yes, they are—but, not the kind of noise that stems from the imperfection of real instruments, the bane of any experimenter (except of course when they run their tests in Gedanken only, the approach favored and pioneered by Einstein). The wrinkles in question are *not* an instrumental artefact, they are a soft clatter that the cosmos itself has produced and that fills every bit of space in the Universe. For a moment, imagine that the expansion after the Big Bang occurred in all directions under exactly the same conditions. No early fluctuations—quantum in nature or otherwise. In that scenario, the map in figure 6.3 would show only a single color, namely that of the prevailing average temperature. All points of decoupling between light and matter would have happened simultaneously and they would lie exclusively on a single sphere, a single event horizon. By what mechanism, then, could the temperature have changed, even if by the tiny amounts we measure? The brief inflationary episode is over and particles are condensing out of the primordial soup. Before the cosmic microwave radiation was unleashed, light interacted strongly with the electrically charged particles—in the later phase mostly free electrons—that filled space. Therefore, if the particle density exhibits local fluctuations, however miniscule, light decouples earlier (fewer electrons) or later (more electrons) in that region compared to the overall average. But if light is freed earlier or later, the ever-present expansion of the Universe has left the ambient temperature a bit hotter or colder, respectively, thus causing the temperature fluctuations observed today. Viewed this way, the faint ripples in the cosmic microwave background mark the beginning of the large- and fine-scale structure of all the matter in the Universe we observe today, almost 14 billion years later. Detailed models of first stellar and then galactic evolution are consistent with the cosmic age determined from the Hubble flow. What is even more telling is the fact that when these same models fast-forward the spatial pattern of the microwave background, they predict very well the currently observed angular size distribution of matter in the universe, from galactic clusters to the foam-like filaments along which cosmic matter seems to be organized.

We have already used the raisin bread analogy in the context of the expansion of the Universe, with the raisins playing the part of galaxies. However, for the very early phase of the cosmos, a more apt analogy for what was occurring is that of a soup—a very finely blended, almost perfectly homogenous soup. So where do the lumps come from? How does the soup coalesce, first into the microscopic constituents of matter and then into the large scale spatial structure we observe today?

6.4 Supporting evidence 2—it is elementary

'We are stardust,' proclaim Crosby, Stills, Nash & Young in their song *Woodstock*. They are literally right, we are made of stardust. 'We are billion year old carbon', they go on. Right again. The carbon, nitrogen, phosphor, calcium, iron and—apart from hydrogen—any other element in our bodies and everywhere else in the solar system are the ashes of nuclear fusion processes that were burning inside the ancestor star (or, more likely, stars) that gave birth to the Sun and its planets. For most of their existence, stars keep a self-regulated size due to a balancing act between the gravity that pulls the stellar material inward and the hellish temperature produced by nuclear burning in the star's core that effectively pushes matter outward. It all starts with hydrogen, which is abundantly available from the earliest moments and is the most efficient nuclear fuel there is. Eventually, though, the star runs out of fuel. When that happens, and what unfolds next, depends on the size of the star. Running out of gas with your car means only one thing: your car stops. Stars, on the other hand, have some tricks up their sleeves, or rather inside their bellies. Roughly speaking, stars can recycle the new, heavier elements produced in previous fusion cycles, using them as input for a subsequent run of energy production. It is as if cars stored their own exhaust gases and used them as fuel once the tank is empty. The fusion of hydrogen, for example, leads to helium, which can combine to carbon and so on, admittedly along a somewhat complex pathway. However, compared to lighter elements, the fusion of heavier elements—into even heavier ones—requires higher temperatures and/or higher density to get started. That is where gravity comes in handy. Once one type of fusion terminates because of lack of fuel, the temperature drops and gravity gets the upper hand. As a consequence, the star contracts, or at least the inner parts do. This collapse releases gravitational potential energy, which heats up and compacts the core. Whether or not this leads to a new round of fusion depends on how much temperature and density increase occurs, which in turn depends on size—the more mass contracts, the more gravitational potential energy is released, the hotter the core gets. If the star has enough mass left at that stage, at least about twice that of the Sun, several more rounds of cooling, contraction and reignition ensue, sometimes accompanied by intense light production, rendering the star many times brighter than our Sun, for example.

The very brightest light shows are on display when a massive star has developed an iron core. Then fusion has run its course, since the combination of two iron nuclei no longer releases energy. As the temperature drops, gravity is the winner again, squeezing the material of the core tighter and tighter. How tight you ask? It is a really good squeeze. Gravity relentlessly increases in strength in direct proportion to mass. If enough mass is present, gravity can even push electrons into protons, converting them into neutrons—almost the time-reversed version of neutron beta decay discussed in the previous chapter. In any case, the process of contraction releases an enormous amount of gravitational potential energy that heats up the outer layers of the star, where new fusion cycles are ignited and even more elements are created, including silver and gold[10]. Because the new energy production turns on quasi-instantaneously over a large volume, it ignites an explosion of cataclysmic proportions, complemented by one of the brightest light shows you can imagine [9]. A supernova appears in the sky, bright enough to be seen with the naked eye. While modern telescope-based observation has yielded a copious database of supernovae, only a handful of such events have been seen by eye in our galaxy during the past 1000 years. The last one (SN 1604), described in detail by Kepler in 1604, was so bright that it was visible during daytime for several weeks. Only 32 years before, Tycho Brahe had coined the name *stella nova*, Latin for *new star* when he wrote about 'his' supernova (SN 1572) in the constellation Cassiopeia. On the one hand, it was an obvious and apt term, but on the other, it is quite ironic, since *stella moriens*, as we now know, describes the case more accurately. Then again, in the evolutionary context, death and new life are not that far apart from each other. One of the earliest recorded events (SN 185) dates back to the year 185, when Chinese astronomers mention a mystifying 'guest star' that appeared out of nowhere and remained visible for eight months. According to a corresponding Wikipedia entry [10], the observers described the image in the sky as looking like 'a large bamboo mat' and displaying 'the five colors, both pleasing and otherwise.' Hmm. It is very likely that the filamentary structures in the left panel of figure 6.4, a composite of four space telescope images[11], are remnants of this same transient event. Not much is left from the mat, and the colors have faded as well, since what you see are false colors, added to show which satellite telescope saw what part of the structure. The detected light itself is in the X-ray and IR part of the spectrum, invisible to the human eye. Many other examples show that supernovae do indeed scatter the seeds from which a next

---

[10] Recent research [8] indicates that these two precious metals, as well as certain isotopes of other elements, can be traced to very specific types of supernovae.

[11] X-ray data from NASA's Chandra and ESA's XMM-Newton observatory were converted to the blue and green colors. Infrared data from NASA's Spitzer space telescope and Wide-Field Infrared Survey Explorer (WISE) are shown in yellow and red.

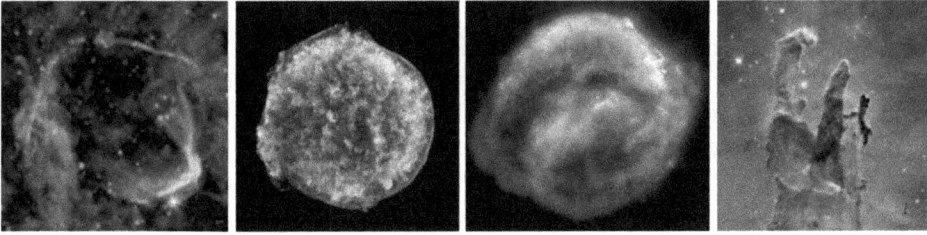

Figure 6.4. Hubble images of the supernovae mentioned in the text, in chronological order from left (SN 185, SN1572, SN1604) plus pillars of creation. Courtesy of NASA.

generation stars and their companions are born. Images from the Hubble space telescope, like the pillars of creation also reproduced in figure 6.4, vividly illustrate regions of dense dust and a high rate of star formation.

In summary, the flare-up accompanying a supernova scatters the outer layers of the old star all over the neighborhood, providing the raw material for the formation of comets, asteroids, planets—you name it—and whatever happens afterwards in, on and around these planets, including of course Earth. Almost all chemical elements can be traced to processes, either fusion or slow neutron capture, occurring during the lifetime of the star. Only a few very heavy elements, such as uranium, are thought to have been produced in the supernova event itself. Our current best models of star evolution after the Big Bang, coupled with a detailed understanding of the nuclear processes that make stars shine, yield the elemental composition of matter in our solar system in good agreement with observation.

So far, so good. But where does the helium and hydrogen come from that must have fueled the very first generation of stars? No prior star existed that could have produced helium, while hydrogen is a basic substance that cannot be obtained by *any* fusion process. To paraphrase Sherlock Holmes: elemental—the only logical conclusion is that hydrogen and at least most of the helium must have been around *before* the first star ignited its nuclear fire. Isotopes of these two elements, together with a very small trace of lithium, were bred in the first few minutes of the life of the cosmos. In that short period of time, conditions were just right for quarks, the basic building blocks of ordinary matter, to condense first into protons and neutrons and then into deuterons, the nuclei of the light isotopes of hydrogen, and finally into helium. Given the pace of reactions, the value of the neutron lifetime (see section 5.4) and the conditions at that time, a number ratio of about 92:8 in favor of hydrogen or, equivalently, a 24% helium mass fraction can be predicted, which is in line with the observed ratio of hydrogen and helium all over the Universe. There is even more. A detailed analysis of reactions that occur between the primordial building blocks present between about one second and one minute or so after the Big Bang—mostly neutrons, protons, electrons, anti-electrons (positrons), photons and neutrinos—predicts the relative abundances of the light elements up to lithium, which depends strongly on the density of ordinary matter relative to photons. What is missing to test the calculation, is a cup of the cosmic soup as it existed in the early stages of the universe. Extensive stellar burning has altered the initial values so much that the present elemental abundances that could be measured easily in our cosmic neighborhood do not provide a meaningful comparison. Thanks to a lucky circumstance and clever thinking—and the well tested method of spectroscopic fingerprinting—astronomers have been able to identify several classes of objects that at least approximate the primordial state of affairs. For example, by observing stars or galaxies with systematically varying content of oxygen and nitrogen—elements that could only have been produced by stellar fusion and not in the Big Bang itself— the zero N and O case can be extrapolated. In this and other manners, the primordial abundance of the two helium isotopes ^{3}He and ^{4}He as well as heavy hydrogen and ^{7}Li has been determined. Overall, the agreement with the predicted values (see figure 6.5) is remarkable, even if in the case of lithium some discrepancies remain.

Thus, the coherent story of how the chemical elements came about provides a third argument bolstering the Big Bang idea. While the superhot mix expands and cools, morsels appear out of almost nothing. First, basic ingredients emerge in the primordial soup, which is still a bit bland. But soon all the necessary components emerge to turn the broth into a delicious, complex bouillabaisse. And we have figured out the recipe.

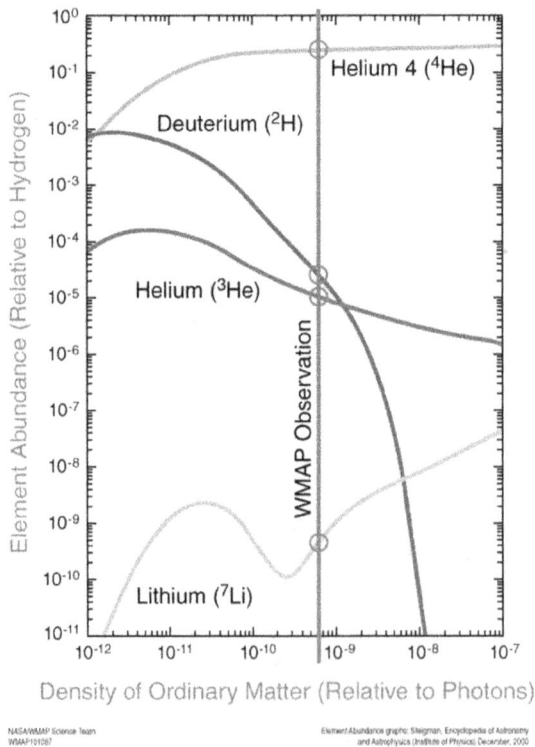

Figure 6.5. Predicted dependence of light element abundance on the density of ordinary matter in the early universe (adopted from the homepage of the Wilkinson Microwave Anisotropy Probe mission [11]).

6.5 Supporting evidence 3—it is dark at night

Everyone knows that it is dark at night. And yet there is a lesson to be learned from this obvious fact that you may not yet know. If there is an infinite number of stars, and if the Universe has existed forever, there should be stars everywhere you look. Not the tiniest patch of the night sky should be left in the dark. Here is a quick proof. Suppose that, at least in the grand scheme of things (after all, we are talking infinity), stars are sprinkled uniformly throughout the Universe with a small but non-zero density. Further imagine that the Universe is a huge onion, with you in the center. In each of the infinitely many layers of the onion, the number of stars grows as the area of the onion shell, i.e. as the radius squared. On the other hand, the intensity of the stars dims as the inverse of the radius squared. Therefore, each layer contributes—on average—the same small, but finite intensity to the total brightness at the center. Since, as assumed, the number of layers is infinite, the converging light will be infinitely bright. Even closing your eyes would not help—the night sky would not be dark. There are two ways out of this conundrum, also known as Olbers' paradox[12]: either the Universe is not infinite or it has not existed forever. Already the philosopher Immanuel Kant had speculated about what the visible pattern in the night sky is due to. He proposed [12] that the stars and Milky Way are part of an island in the Universe and that we see this island from within [13]. Or put differently: we are in a forest of burning candles (minus the sticks) that are arranged in a disc-like pattern around us. In certain directions, we look towards the thick of the forest and in others, we see past the edge into the neighboring void. When viewed through a telescope, some of the light specks were already known at Kant's time to be fuzzy ellipses, quite distinct from the images of stars. Kant speculated further that these blurry objects are other islands in the Universe or, as we would call them today, other galaxies. Kant was right. And he was also smart. Later, in his *Kritik der Reinen Vernunft* (*Critique of Pure Reason*) [13], published in 1781, he recanted and argued that questions of cosmology, such as whether the spatial or temporal extent of the Universe is finite or not, lack a solid basis to be answered and thus are, in a strict sense, meaningless. Well, at his time they were. In and by itself, Olbers' paradox is indeed not proof of cosmological finiteness. However, it is true that a dark night sky is consistent with the Big Bang theory. And that is all we can ask for—remember that the scientific method only allows theories to be falsified, never to be proven.

6.6 Standard candles and a very long ladder

One recurrent theme of the book is that ideas are important, but so are the experimental and observational tools and methods that test these ideas, and occasionally even lead to new ones. The Big Bang theory is no exception. In that

---

[12] Heinrich Wilhelm Olbers (1758–1840), a daytime physician and nighttime astronomer who practiced both of his callings in Bremen, Germany.

[13] The subtitle of Kant's early work [12] reveals his ambition: *Attempt on the Constitution and the Mechanical Origin of the Entire Universe, treated according to Newtonian Principles* (*Versuch von der Verfassung und dem mechanischen Ursprunge des ganzen Weltgebaudes, nach Newtonischen Grundsatzen abgehandelt*).

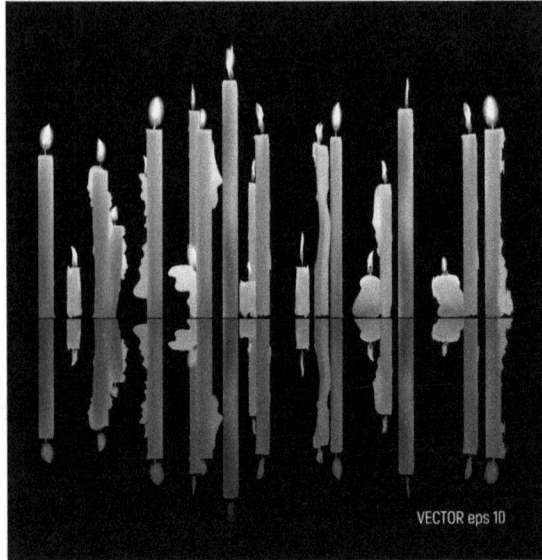

Figure 6.6. Olbers' paradox—given enough stars, the night sky should be bright. Source Alexey U/Shutterstock.

spirit, a few more remarks are in order about the ways the central ingredient, the Hubble law, was established and modified. In doing so, we will also be reminded of the tight connection between space and time. The material in this section covers an enormous range. It can and does fill entire books that are significantly longer then this slim tome. So, if you are intrigued, puzzled, or unsatisfied—keep looking.

Figure 6.1 illustrates how in one specific survey the redshift of galaxies varies with their distance from us. Establishing the redshift is a quite straightforward business in principle: with your telescope pick a galaxy, direct its light into a spectrometer (see chapter 3), and compare the resulting spectrum with that of a suitable light source in the laboratory. If the galaxy has no radial velocity relative to us, all of the emission and absorption features will be located at the wavelengths found in the laboratory reference spectrum. If there is radial motion, lines will be displaced—to the red (large wavelengths) for receding sources and to the blue (shorter wavelengths) for approaching ones. More quantitatively, the Doppler effect results in a wavelength change $\Delta\lambda$ that is equal to the product of wavelength λ of the same line from a stationary source and the speed v in relation to the speed of light c, i.e. $\Delta\lambda = \lambda \cdot (v/c)$. This gives us a consistency check when more than one line is present, which is usually the case. Of course, the challenge is to apply this technique to sources that are far away. Light spreads out in a spherical shell from the source and thus its intensity diminishes in inverse proportion to the total area of the shell. Since the latter increases as the square of the radius of the shell, light intensity decreases as the inverse of the distance squared of the light source from the detector. For example, if a star is detected from our neighboring galaxy Andromeda, the light has traveled about 2.5 million light years, a distance that is approximately 25 times larger than

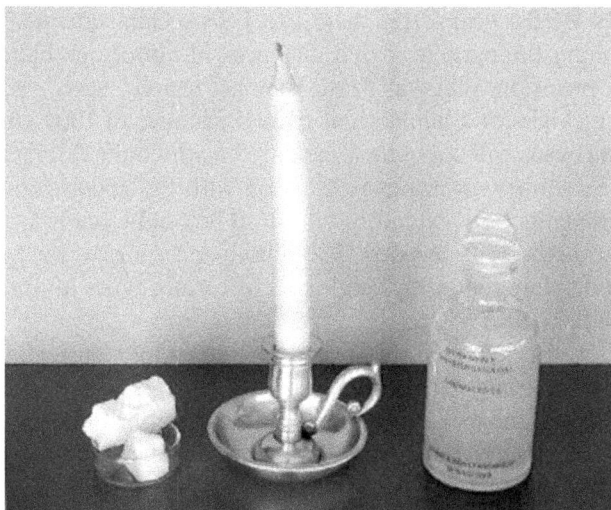

Figure 6.7. Standard candle—same composition and make implies same brightness. This image has been obtained by the author from the Wikimedia website where it was made available under a CC BY-SA 3.0 licence. It is included within this article on that basis. It is attributed to Genevieve Anderson.

that of a similar star at the opposite end in our own galaxy. Thus, the Andromeda star would appear dimmer by a factor of about 625. And that is just the beginning. The furthest objects we can see are of the order of 10 billion light years away from Earth, which corresponds to a decrease factor, again compared to the star across our own galaxy, of also ~10 billion[14]. It seems hopeless to try to detect such faint light, and indeed we cannot observe ordinary stars over such vast distances. However, individual objects exist—type Ia supernovae—that shine so brightly they are visible across the entire universe. Because these particular supernovae have a consistent peak luminosity, they can be used as so-called *standard candles*. If you compare two such light sources and one is four times dimmer than the other, you immediately know that it is also twice as far away. Using several different types of standard candles, astronomers have produced a distance ladder whose individual rungs are calibrated and which span distance scales ranging from those relevant for the solar system, close stars and the Milky Way, all the way to the edge of the Universe.

The first step in measuring cosmic distances is the plain old triangulation method we encountered earlier (see, for example, chapter 4). Usually the triangle consists of Earth's orbital diameter as the baseline, with the star in question at the apex. Determination of the two angles between baseline and the direction to the star measured at the two endpoints of the baseline, together with the known length of the baseline, allows one to obtain the distance to the star. Of course, you need to be patient (and careful), since the two angular determinations must be separated by six months. The distance to a handful of stars had already been established in this

---

[14] That is just a coincidence. The arithmetic goes like this: ratio of distances (in light years) = $10^{10}:10^5 = 10^5$. Thus, the brightness changes by factor $(10^5)^2 = 10^{10}$.

fashion by 1830. By the year 2018, thanks to ESA's Gaia satellite, we will have a database containing the parallax-based distances of about one billion stars in our galaxy with a precision that—at least for the nearest stars—is equivalent to measuring the thickness of a human hair from a distance of 1000 km. This massive and reliable star census will strengthen the base on which the distance measurement ladder with all its subsequent rungs rests. Even with the upcoming improved data, parallax measurements only extend to about 30 000 light years. In order to reach farther, different methods are needed. The subsequent rungs of the ladder are based on a detailed understanding of the brightness of certain types of objects we see, i.e. on standard candles.

Stars vary a lot, they come in different sizes, colors and ages, and thus they also differ a great deal in terms of their brightness. If two stars vary by a factor of two in luminosity, that could indicate that the dimmer one is four times further away than the brighter one. Or they could be at the same distance from us and simply be different star types. There are ways to classify stars, mostly based on spectroscopic studies of the starlight, that give us a fair idea of whether or not two stars ought to have the same brightness. Still, this approach to extract distances is tricky and full of pitfalls. It is therefore very helpful that the so-called Cepheid variable stars exhibit a unique relationship between their peak brightness and the period with which they pulsate. Since it ranges from days to months (see figure 6.8), the period of variability can be measured easily. In 1908, Henrietta Swan Leavitt discovered this property by concentrating on variable stars in the Large and Small Magellanic Clouds, two nearby galaxies. Because the size of *any* galaxy is much smaller than its distance to us, *all* its stars are roughly equidistant to us. Even for the closest galaxies, the distance varies by only a few percentage points. Thus the apparent brightness of the variables that Swan Leavitt observed is scaled by the same factor, namely the inverse distance squared. Comparison of the apparent luminosity of variable stars with the same period in the Magellanic Cloud and in our own galaxy, as well as with other stars at known distances, allows calibration of the Cepheid method. It is this approach that Hubble and others used early on. With effort and skill, these

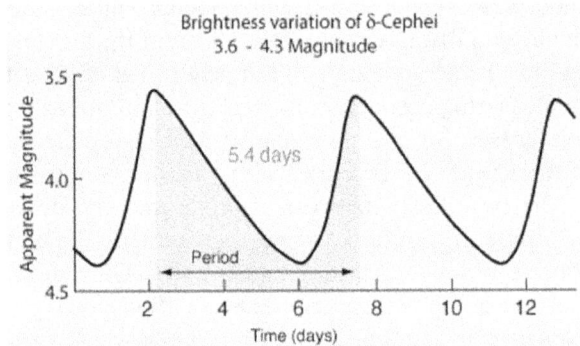

Figure 6.8. Schematic intensity variation of Cepheid (adopted from an entry in Hyperphysics—http://hyperphysics.phy-astr.gsu.edu/hbase/Astro/cepheid.html. Reproduced with permission from Carl R (Rod) Nave, Georgia State University.)

observers were able to push out the distance boundaries to several tens of mega parsec, roughly 100 million light years.

But that is about all one can do. As with any other source, Cepheid stars dim with increasing distance. We need yet another, brighter source that can serve as a standard candle. In the segment on nucleosynthesis, we mentioned that supernovae events, the death throes of massive stars, go along with intense firework displays. Indeed, supernovae can shine more brightly than their entire host galaxy. That is useful for our purpose, but supernovae do not come in a standard size and their brightness can vary by several orders of magnitude. Except for one very special category, unglamorously called type Ia supernovae. If a dense white dwarf—the corpse of a small star—has a stellar companion, things can get interesting. Dual-star systems are not at all rare. Remember that Doppler's motivation, back in the mid-19th century, was to elucidate certain properties of binary stars. Statistics dictate that, in a certain fraction of such systems, the partner star has a large mass and is close enough for the gravitational pull of the white dwarf to suck material from its outer atmosphere. As long as its mass is insufficient to further compress the cold carbon core, the white dwarf remains cold and dormant. But by feeding off its large companion star, the little one grows fatter. And fatter. And one day, the now former white dwarf reaches a critical mass that allows the gravity to overcome the quantum mechanical forces that keep the electrons apart with the consequences described above. A supernova is triggered. Because type Ia supernovae reach criticality by mass growth from below and towards a sharp threshold, all events of this type have the same mass. Thus, they all have the same intrinsic brightness. They are the perfect standard candle. But how would you recognize if a bright new star appearing in the night (or even day) sky is of type Ia? That is possible because all such events start with the same composition, the remnants of a small star that only burns hydrogen and helium. Consequently, the brightness of type Ia supernovae varies in a characteristic and reproducible manner. For example, it takes about 20 days after explosion to reach maximum luminosity. It is as if all the organizers of Fourth of July fireworks in the US agreed to use the same number of shells, with the same chemical composition, and launched them with the same choreography. With all these advantages, type Ia supernovae have made a tremendous impact on cosmology. A sizable fraction of the early observation time of the Hubble Space Telescope was dedicated to a systematic survey and careful measurement of these standard candles. Among the most distant objects are explosions with a redshift corresponding to a distance of about ten billion light years. Even at these tremendous distances, a thousand times further than the most distant objects Hubble had originally analyzed, these faint galaxies conform to the Hubble law—almost. There is a small but significant deviation, which we will discuss in the next section. One interesting detail that convinced pretty much everyone about the reliability of the SN Ia method is the fact that the observations are consistent with special relativity. At one billion light years, the observed redshift corresponds in round numbers to a velocity of 21 000 km s^{-1}, about 7% of the speed of light. We have seen that moving clocks run more slowly and that the light curve of a type Ia supernova follows a predictable pattern. Although the details are complex, a careful model [14] of the observed light curves of 13 high-redshift supernovae—the most distant and hence fastest moving away at about 40% of the speed of light!—clearly reveals that the faster the supernovae move away from us, the slower the processes unfold, in complete agreement with relativistic time dilation (see chapter 4). Not only does this consistency strongly support the use of type Ia supernovae as standard candles, it also rules out Zwicky's alternative explanation of the galactic red shifts. He had speculated that the Universe is static and that, on average, distant sources are stationary. In his view, the observed redshift is due to scattering of light. In such 'collisions' of photons with whatever fills the vast space between the emitter and the detector (aka cameras and spectrometers at the end of our telescopes) energy is lost. Hence the frequency of photons, being proportional to their energy, is reduced or the wavelength is increased (see chapter 3). However, this 'tired light' scenario, as it was often called, does not explain the change in the observed light curves. Stationary sources should have the same time dependence, even if the spectrum might be red shifted.

In summary, interlacing the various methods of distance measurement—parallax, brightness of spectroscopically well-characterized stars, Cepheid periods and brightness of type Ia supernovae—allows astronomers to determine the distances of light-emitting objects extending to the furthest reaches of the observable universe. Start with triangulation. Classify the stars with known distances according to their emission spectra. From a sufficiently large sample, correlate intrinsic brightness with star type. Then the measured, apparent brightness of stars associated with a unique type can be used to infer their distance. Variable Cepheid stars offer the additional property of one-to-one correspondence between brightness and periods. Finally, type Ia supernovae, whose observations currently number in the tens of thousands [15], are so bright that they can be detected even at the edge of the observable universe. While rare, some have been seen in galaxies that also host Cepheid stars and a few cases also exist where two such light flashes occurred in the same galaxy (with both flashes showing the same brightness). Thanks to modern instruments, with the Hubble Space Telescope having led the way, and efficient search strategies, the scope and quality of the available redshift data have turned cosmology into a precision science. One of the already-mentioned results is the impressive consolidation of the Hubble constant. As was the case for Earth, finding the age of the Universe followed a long and winding road and, also like the case for Earth, it arrived at a value that was substantially older than was assumed early on.

6.7 Was there anything before the Big Bang?

It is natural to wonder what lies behind the horizon—it might be useful, it might be dangerous, or it might be beautiful. But maybe there is nothing there at all. If the Earth is a flat disc, it must have an edge. Unless it is an infinite disc. But then the Sun has no place to sink below the horizon. Good point. So, the Earth must be of finite size. But what lies beyond the edge? Water, water everywhere—as far as you want it to reach, with Earth swimming on top. That way, at the end of each day, the Sun takes a bath. All this is possible until the evidence says otherwise. So, what about the horizon in time, the Big Bang? What was there before it? Surely something. How

Figure 6.9. What lies behind the horizon has always intrigued curious people.

could our universe have come into existence out of nothing? And what about the future? Will the world end? How will it end? And what lies beyond the end? If any answers are offered, they differ a lot depending on whether they are inspired by religion, philosophy, or even plain storytelling. But can science contribute anything? As always, science is constrained by its own methodology. If a hypothesis cannot be tested, it is not a scientific hypothesis. On the other hand, it is quite surprising in what subtle ways a theory has falsifiable consequences. As long as a mathematical description of nature has not been proven incorrect, it is legitimate to take all its predictions at face value. Indeed, that is one way to expose the theory to scrutiny and possible contradictions. So, what does current cosmology have to say about the very origin and possible fate of the Universe? As it turns out, not all that much on the first point, but it does give a few detailed options for the latter.

Blame it on a fluke that we are here, or rather that anything at all is here. A fluke like in quantum fluctuation, which is the physicist's way to wave a magic wand. While it is permissible to be a bit more specific than that, to say a whole lot more would be misleading—at least I think so. In chapter 3, we encountered the Heisenberg uncertainty relations, which encapsulate the difference between the certainties of Newtonian physics and the probabilities of quantum mechanics. One of these relations asserts that the energy content associated with an event of duration Δt is undetermined to within $\Delta E \geqslant h/\Delta t$, where h is the Planck constant. The briefer the duration, the more energetic the quantum fluctuation can be. In a complete vacuum, for example, with nothing at all in the neighborhood, a particle with relativistic rest mass energy mc^2 and its antiparticle can come into spontaneous and fleeting existence for a duration $\Delta t \approx h/2mc^2$. For a pair consisting of an electron and a positron, that duration is only about $4 \cdot 10^{-21}$ s. Nevertheless, this temporal morsel is only about a factor of 1000 shorter than the length of processes at the atomic level that researchers can currently create and control [16]. And compared to the length of the inflationary epoch of the Big Bang, $\sim 10^{-32}$ s, an interval of 10^{-21} s is a downright eternity. There is an even shorter time span than cosmic inflation that has meaning in physics. Planck observed that the gravitational force constant G, the speed of light c and the Planck constant h can be combined to yield a unit of time. Known as the Planck time t_P, the expression $\sqrt{hG/c^5}$ is the simplest expression that fits the bill, yielding a value of roughly 10^{-43} s. Given the intimate connection between spatial and temporal dimensions (see chapter 4), it should not be a surprise that there is also a Planck length ℓ_P, and that it equals the distance light travels in one unit of Planck time, about $4 \cdot 10^{-35}$ m. There is no universally accepted interpretation or significance for these two parameters. However, because they contain the physical constants that pertain to gravity (G) and quantum physics (h), it is reasonable to assume that Planck time and length set the temporal and spatial scales at which the gravitational force acquires a quantum nature. Because we do not yet have a theory of quantum gravity, it also seems reasonable to me that physics cannot be on firm ground when we approach the Big Bang to within one Planck time or so. In any event, even the Planck time is not short enough for a 'fluctuation' to account for the mass and energy content of the Universe. In short, we can only speculate, but we really do not know what set off the Big Bang.

6.8 Will it end?

While the exact origin and nature of the Big Bang are murky, we have a clearer picture of the possible future fate of the Universe. Given the current *modus operandi* of the Universe, there are two logical possibilities: either the cosmos keeps expanding forever or it will, at one time in the future, reverse direction and contract. There are different ways in which each of the two scenarios might unfold, but basically those are the choices. Without thinking too much about it, we might be inclined to bet on the second case. There is no avoiding gravity, the 'tyranny of weight' as it is called in a book on transporting sculptures in renaissance Italy [17]. Or, as the saying goes, 'what goes up, must come down.' Unless it does not. Throw up a ball and it will return to ground after reaching a certain height. Throw the ball faster and both its 'hang time' and maximum elevation above ground increase. You do not have to throw the ball infinitely fast (let alone that you cannot) for the maximum height to be infinite and the ball to be gone forever. The simple reason for this observation is the fact that the gravitational force diminishes in strength with increasing distance from the force center. All satellites and space vehicles launched to the Moon and beyond are testimony to the fact that with a certain threshold speed and above there is no return to ground—the ball, the satellite, the deep space probe will *not* come down. For a launch from Earth's surface, the so-called escape velocity comes out to be a little more than 40 000 km/h[15]. This value increases with the mass you are trying to escape from. Roughly the same arguments can be applied to the

---

[15] At that speed, a trip to the Moon would take a mere 9.5 hours. The Apollo missions to the Moon had to take a different strategy. There is no practical—and safe—way to launch a space vehicle from Earth with such a high speed at once. A prolonged phase of low acceleration is used instead. Also, the above discussion contains several simplifications. For example, the influence of the rotational motion of the Earth has been ignored. Still, the essential idea is valid.

whole universe. Pick a random point in space and outline a huge sphere centered at the point you chose. If we assume that the Universe is isotropic and homogeneous, at least at large scales, the specific location of the point and the specific radius of the sphere do not matter, provided that the latter is larger than the characteristic length scale associated with any 'lumpiness' of the Universe. With that term, I refer to the raisins in the dough: the galaxies and clusters of galaxies into which matter clumps. In other words, the sphere must be humongous, say tens of millions of light years in size. Fine. Since the size of the visible universe measures into billions of light years, there is still plenty of universe outside the sphere. And that part is moving away just as our tossed ball is. And it will return just as the tossed ball if its speed is not above the escape velocity. But what is the escape velocity here? It is a bit more complicated, since we are now dealing not with a single object but with a spherical distribution of matter. On the length scales employed here, it is okay to think about the Universe as being homogenously filled with matter at an average mass density ρ. The total mass in the sphere is then simply the product of this density and the volume of the sphere. Analyzing what will happen to a thin shell just outside our big sphere turns out to be all we need to find out what will happen to the whole structure. From its location at a great distance R from the center, we know that the shell is part of the Hubble flow and that it has a speed away from the center given the Hubble law, i.e. H_0R. If the gravitational potential of this shell is smaller than its kinetic energy, then the shell moves faster than the corresponding escape velocity and will never return. By the same token, if the gravitational potential is larger, the current velocity is not sufficient to overcome the long reach of the gravitational pull. And if the two energies should be just equal, the shell will come to rest—but only after an infinite amount of time and an infinite distance. This last case occurs, for a given value H_0 of the Hubble constant, for a well-defined critical density ρ_c that can be derived with first-year college physics. Which is astounding, since supposedly cosmology is based on general relativity, which is definitely *not* first-year college physics material. In physics, it sometimes happens that simple arguments (simpler than one might expect to be justified) still capture the essence and yield a correct result. The equality of the two forms of energy applied to the outer shell lead to the following valid expression for the critical density

$$\rho_c = \frac{3}{8\pi}\frac{H_0^2}{G},$$

where H_0 is the current value of the Hubble constant, about 70 km s^{-1} MPc^{-1}, and G is the gravitational force constant, $6.67{\cdot}10^{-11}$ N·m^2 kg^{-2}. With these two values, the critical density comes out to be about $9.2{\cdot}10^{-27}$ kg m^{-3}, or about five hydrogen atoms per cubic meter. What does that mean? It simply implies that if a cubic meter of space, suitably averaged, contains more than five hydrogen atoms, the rate of expansion will slow down and reverse, and all that went 'up' will come crashing 'down'. In that case, the Universe will steer towards what has been dubbed the Big Crunch. So, how many atoms are there in space? Will there be a crunch? Since the average must be taken over a sphere that is large enough to contain many galaxies, we need to know how many galaxies we caught with our spherical net, how many stars a typical galaxy accommodates and how massive stars are. Once we know all that, we can calculate the actual mass density from observable quantities. Easier said than done, but doable.

Our current best estimate [18] for the mass density of ordinary matter in the Universe is a paltry 0.25 hydrogen atoms or so per cubic meter, clearly not enough for closure. Since the mass density is so low, the Universe can be considered to be almost empty, so that the rate of expansion should stay approximately constant over the history of the Universe. But wait, there is more, literally more. Apart from the material we are familiar with and from which stars, planets, interstellar dust and black holes are made, the Universe contains another kind of matter, dark matter. What this stuff is, we do not yet know. But there is clear evidence that it is present in abundance all over the Universe, and that it is gravitationally active. The existence of dark matter was inferred as early as 1933 by Fritz Zwicky. All galaxies and also clusters of galaxies rotate around their intrinsic center and gravity holds the objects together. In all cases, though, the mass inferred from the luminosity of these objects is much too weak to do the job. Dark matter is thus invoked to keep galaxies from shedding their stars. An additional indication for the existence of dark matter comes from the now commonly observed bending of light by galaxies, the so-called gravitational lensing, which is always much stronger than the observed ordinary mass of the light-deflecting galaxy would indicate. Based on all such observations, we presently infer that the Universe contains five times more mass bound up in dark matter than all the atoms and molecules combined that make up the stuff you learned about in school. It is quite amazing that we can be so precise about something we know hardly anything else about. But even if we do not know what exactly dark matter *is*, we know a bit about what it *does*, namely exerting gravitational pull on everything, including light and ordinary matter. Thus, we can add dark matter to the tally for the critical density, which now comes out to be equivalent to the mass of about 0.25 (ordinary matter) plus 1.25 (dark matter) for a total of 1.5 hydrogen atoms per cubic meter. Still shy of five, the number we need for the Big Crunch, and still low enough for the Universe to be in near coasting mode. Near coasting, since the presence of matter—dark and otherwise—should have an observable effect on the rate of expansion of the Universe: its gravitational pull should slow down the expansion. Therefore, if we look back in time, as we do when looking at distant galaxies, we should find a ratio of redshift to distance that is lightly higher than the current value (which is found for 'nearby' galaxies).

Until about the late 1980s, the available galactic redshift data indicated that the Hubble constant comes out constant throughout cosmic history. The statistical and systematic error bars were large enough to hide the expected slow-down in expansion rate. Pleasingly and reassuringly, the observational data agreed with Lemaître and Friedman's solutions of Einstein's field equation of gravity. Leaving the unsettled business of the nature of dark matter aside, all seemed to be in consistent and perfect order. It appeared that the modern saga of the world was complete and that its last chapter had been written.

Therefore it came as a shock—and was greeted with much skepticism—when two research groups announced that they had found compelling proof for an

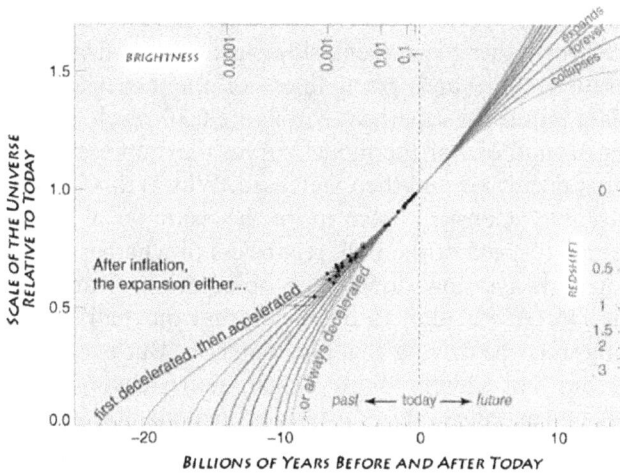

Figure 6.10. Dark energy. The brightness of type Ia supernovae indicates their distance and therefore how long ago they exploded. Their redshift indicates how much the universe has expanded since then. Together, and with enough samples, their brightness and redshift indicate the expansion history of the universe. Researchers expected to find that the expansion was slowing, or at least coasting (red lines). Instead they found that it was accelerating (blue lines); expansion will probably continue to accelerate in the future. Reproduced from [19] with the permission of the American Institute of Physics.

acceleration rather than a slowing of the cosmic expansion rate. There had been earlier hints that not all was well with the narrative outlined above, but since the researchers involved in the new findings were fully aware of the implications, they hesitated to formally announce the new discovery until 1998. It helped that the two groups arrived independently at the same result: starting at around 7 billion years after the Big Bang, the rate of expansion of the Universe increases! The careful analysis of many very distant type Ia supernovae, the standard candles introduced above, seems to leave no room for any other interpretation. Figure 6.10 summarizes what we now know. In principle, the graph shows the same kind of data that Hubble had championed (see figure 6.1): the redshift of an object on the vertical scale versus its distance on the horizontal axis. However, there are some variations in the way the data is presented. First, spatial distance is converted into temporal distance in the usual manner. Light from a galaxy that is one million light years away was launched one million years ago. With the zero point referring to the presence, past and future are designated by negative and positive number of years, respectively. For obvious reasons, no observational data, but only calculated projections, are shown for the time beyond the current epoch. As was explained above, distance determination using standard candles involves the comparison of their measured brightness with their brightness at a standard distance. This explains the conversion between the top and bottom horizontal axes. Likewise, redshift on the right vertical axis is related by the Doppler effect to the speed at which a given galaxy is receding. This indicates how the right vertical axis of figure 6.10 is obtained in principle. One more conversion, on the basis of Hubble's law, turns the red shift to a cosmic length scale with a present day value of one and a start value of zero for the Big Bang. This is the left vertical axis. Several curves delineate model calculations of cosmic dynamics that reflect either permanent slowing down (red and orange lines) or eventual acceleration (blue and green lines) of the expansion of the universe. Observational data points for supernovae in figure 6.10 reach back about 10 billion years and cluster around one of the model curves that represents a universe whose expansion rate first decreased and then increased. Why is this increase in the rate of expansion remarkable? Consider once more the example of the tossed ball. No matter what the initial speed of the ball, regardless of whether it can escape or not, the tossed ball will always slow down. The observation of an acceleration of the expansion of the Universe is akin to observing that our ball speeds up as it rises. Weird. Something must be driving this phenomenon. But even over more than 20 years later, we are not closer to an explanation, leaving room for creative speculation. Only two aspects of this mysterious phenomenon are universally agreed upon: its succinct name of *dark energy* and the fact that it amounts to about 70% of the entire mass–energy content of the Universe. The latter number is derived by varying the dark energy ratio in the corresponding general relativity calculations (the specific model calculation whose corresponding line matches the observation in figure 6.10), which are modified to contain an additional term that is usually called Λ. It is amusing that Einstein had introduced an ad hoc term into his field equations so that they could yield a stationary universe solution—his preferred outcome (see the section on Lemaître above). While he was later quite embarrassed about the introduction of this fudge factor Λ, it now seems that he was right after all, although for the wrong reasons. Like Einstein's cosmological constant, dark energy is supposed to permeate all space uniformly[16]. However, its density is only about 10^{-27} kg m^{-3}, which is equivalent to the presence of less than one proton in a volume of one cubic meter. Therefore, dark energy is utterly negligible as far as any laboratory experiment is concerned. And it also plays no role in celestial mechanics. For example, Earth's gravitational attraction to the Sun is altered by the presence of dark energy in a sphere with a radius of one astronomical unit, the Sun–Earth distance,or about 150 million kilometers. Quite a sphere. Yet this large volume contains dark energy with an equivalent mass of about 140 000 metric tons, equivalent to the mass of two large aircraft carriers. Impressive, but of course completely negligible compared to the solar mass of about $2 \cdot 10^{33}$ kg. Some things just do not matter on human scales, even if they are very important at higher levels of the cosmic ruler hierarchy.

In the scenario that is currently held to be the best match to all available data, we can summarize cosmic history as follows. About 14 billion years ago, all matter and energy that fills our observable universe sprang forth from a singularity. During the inflationary epoch, lasting only the tiniest fraction of a second, space and all it contained expanded in a mad dash by a factor of as large as 10^{26} or so—a stupendous scaling. If the visible universe was the size of a grain of sand at the end of the inflationary fit, the starting size would have been of the order of a half

---

[16] Incidentally, this aspect of dark energy is somewhat reminiscent of the erstwhile ether. Funny, how tenacious some concepts are.

million Planck lengths. During this era, quantum fluctuations imprinted themselves into the fabric of the cosmos, visible in both the cosmic microwave background and the foamy spatial patterns formed by galaxies, clusters of galaxies and clusters of clusters of galaxies. Dark matter plays an important role in this evolution, which also includes a slight slowing of the rate of expansion. At about 7 billion years after the Big Bang, a new form of energy, dubbed *dark energy*, entered the fray and caused an acceleration of the expansion rate. With that and if nothing else were to happen—in my mind a big if—the future of the Universe is sealed. The gravitational pull of dark and ordinary matter is insufficient to stop expansion—even less so with the cosmic engine kicked into overdrive by dark energy. To assume that the book of physics is closed and that all has been discovered is an old sport—Max Planck was told as much when he was a student—without much benefit. Only 20 odd years ago, dark energy entered the toolbox and the vocabulary of cosmology. I think it is a safe bet that it will not remain the last word. In the meantime, it is entertaining to extrapolate from the present state of affairs to what the future has in store for the Universe. The galactic weather forecast calls for dangerously low temperatures.

In *Alice's Adventures in Wonderland* [20], Alice was well advised by the king to begin at the beginning, to go on till the end, and then to stop there. But sometimes it is difficult to tell where in the story you are. We do not know—now or possibly ever—if the wonderland that we inhabit, our universe, even has an end. And should our universe end, contrary to current evidence, this could be the start of another world. Maybe it is true that 'Every real story is a never-ending story' [21].

6.9 The long now

So here we are, about 14 billion years after the cosmic egg hatched, 4.5 billion years after the Earth clumped itself into existence, and an untold, untellable number of years before the Universe encounters its unknown 'final' fate—if there even is one. The boundary between these two staggeringly long periods of past and future is the fleeting now, the present moment. It is all we truly have, yet both past and future are important to us as well. Learning from the past, living in the now and working towards a sustainable and peaceful future are not the worst things we can do. The Long Now Foundation [22], established in 01996, had something like that in mind when they launched their ambitious project of building a large mechanical clock that can last and tell time for the next 10 000 years. Yes, 10 millennia. I really would like to hear it chime.

References

[1] Doppler C 1842 *Über das farbige Licht der Doppelsterne und einiger anderer Gestirne des Himmels* (On the colored light of the binary stars and some other stars of the heavens) (Prague: Borrosch und André)
[2] American Institute of Physics, Center for History of Physics *Spectroscopy and the Birth of Astrophysics*: http://history.aip.org/exhibits/cosmology/tools/tools-spectroscopy.htm
[3] Lemaître G 1927 Un Univers homogène de masse constante et de rayon croissant rendant compte de la vitesse radiale des nébuleuses extra-galactiques *An. Soc. Sci. Bruxelles* **47** 49–59

[4] Friedman A 1922 Über die Krümmung des Raumes *Zeitschrift für Physik* **10** 377–86

[5] Cooperstock F I, Faraoni V and Vollick D N 1998 The influence of the cosmological expansion on local systems *Astrophys. J.* **503** 61

[6] Hossenfelder S 2017 If the Universe is expanding, then why aren't we? *Forbes (online)* (July 28) www.forbes.com/sites/startswithabang/2017/07/28/most-things-dont-actually-expand-in-an-expanding-universe/#f12d96a3d747

[7] Institute of Physics *6 Things you may not know about the afterglow of the big bang* www.physics.org/featuredetail.asp?id=45

[8] Hansen C J, Primas F, Hartman H, Kratz K L, Wanajo S, Leibundgut B, Farouqi K, Hallmann O, Christlieb N and Nilsson H 2012 Silver and palladium help unveil the nature of a second r-process *Astron. Astrophys.* **545** A31

[9] Bennett J, Donahue M, Schneider N and Voit M 2017 *The Cosmic Perspective* (Boston: Pearson) ch 17
Nave R Supernovae module in Hyperphysics: http://hyperphysics.phy-astr.gsu.edu/hbase/Astro/snovcn.html#c2

[10] Wikipedia entry *SN 185*: https://en.wikipedia.org/wiki/SN_185

[11] https://map.gsfc.nasa.gov/universe/bb_tests_ele.html

[12] Kant I 1755 *Allgemeine Naturgeschichte und Theorie des Himmels* (Universal Natural History and Theory of the Heavens) (Engelmann)

[13] Kant I 1781 *Kritik der Reinen Vernunft* (*Critique of Pure Reason*) (Riga: Hartknoch)

[14] Blondin S *et al* 2008 Time dilation in type Ia supernova spectra at high redshift *Astrophys. J.* **682** 724

[15] Guillochon J, Parrent J, Kelley L Z and Margutti R 2017 An open catalog for supernova data *Astrophys. J.* **835** 64

[16] Boyle R 2016 Smallest sliver of time yet measured sees electrons fleeing atom *New Scientist* **11**

[17] Wallace W E 2016 An impossible task *Making and Moving Sculpture in Early Modern Italy* ed K H Di Dio (Oxford: Routledge)

[18] Bennett J, Donahue M, Schneider N and Voit M 2017 *The Cosmic Perspective* (Boston: Pearson) ch 22

[19] Perlmutter S 2003 Supernovae, dark energy, and the accelerating universe *Phys. Today* **56** 53–62

[20] Carroll L 2011 (first published 1865) *Alice's Adventures in Wonderland* (Peterborough, ON: Broadview Press)

[21] Ende M and Doyle G 1983 *The Neverending Story* (Garden City, NY: Doubleday)

[22] https://en.wikipedia.org/wiki/Clock_of_the_Long_Now

Of Clocks and Time

Lutz Hüwel

Bibliography

Part I – A few books on clocks and time:

Arias E F *et al* (eds) 2016 *The Science of Time 2016: Time in Astronomy and Society, Past, Present and Future* The Science of Time Cambridge Massachusetts USA (New York: Springer)

Audoin C and Guinot B 2001 *The Measurement of Time: Time, Frequency and the Atomic Clock* (Cambridge, MA and New York: Cambridge University Press)

Aveni A F 2000 *Empires of Time: Calendars, Clocks and Cultures* (London: Tauris Parke Paperbacks)

Baggott J 2018 *Origins: The Scientific Story of Creation* (New York: Oxford University Press)

Baierlein R 1992 *Newton to Einstein: the trail of light: an excursion to the wave-particle duality and the special theory of relativity* (Cambridge, MA and New York: Cambridge University Press)

Baker G L 2011 *Seven Tales of the Pendulum* (New York: Oxford University Press)

Baker G L and Blackburn J A 2005 *The Pendulum: A Case Study in Physics* (Oxford and New York: Oxford University Press)

Barbour J 2001 *The End of Time: The Next Revolution in Physics* (Oxford: Oxford University Press)

Barnett J E 1998 *Time's Pendulum: The Quest to Capture Time—From Sundials to Atomic Clocks* (New York: Plenum Trade)

Bennett J O *et al* 2017 *The Cosmic Perspective* (Boston: Pearson)

Blanchard A 2006 The big bang picture: A remarkable success of modern science In *1st Crisis in Cosmology Conf. CCC-I* ed E J Lerner and J B Almeida pp 148–59

Callender C and Edney R 2010 *Introducing Time: A Graphic Guide* (London: Icon Books Ltd)

Clegg B 2011 *How To Build a Time Machine: The Real Science of Time Travel* (New York: St Martin's Press)

Coveney P and Highfield R 1991 *The Arrow of Time: A Voyage Through Science to Solve Time's Greatest Mystery* (New York: Fawcett Columbine)

Davies P C W 2002 *How to Build a Time Machine* (New York: Viking)

Davies P C W 2005 *About Time: Einstein's Unfinished Revolution* (New York: Simon & Schuster)

Falk D 2010 *In Search of Time: The History Physics and Philosophy of Time* (New York: Thomas Dunne Books/St Martin's Griffin)

Galison P 2003 *Einstein's Clocks and Poincaré's Maps: Empires of Time* (New York: WW Norton)

Gorbunov D S and Rubakov V A 2011 *Introduction to the Theory of the Early Universe: Hot Big Bang Theory* (Hackensack, NJ: World Scientific)

Greene B 2007 *The Fabric of the Cosmos: Space Time and the Texture of Reality* (New York: Vintage)

Hawking S 1998 *A Brief History of Time* Updated and expanded 10th anniversary edn (New York: Bantam Books)

Hawking S and Penrose R 2015 *The Nature of Space and Time* Revised edn ed *Princeton Science Library* (Princeton: Princeton University Press)

Horwich P 1987 *Asymmetries in Time* (Cambridge, MA: MIT Press)

Jespersen J and Fitz-Randolp J 1999 *From Sundials to Atomic Clocks: Understanding Time and Frequency* (Mineola, NY: Dover)

Kaku M 2016 *Hyperspace: A Scientific Odyssey Through Parallel Universes Time Warps and the Tenth Dimension* (Oxford: Oxford University Press)

Kuhn T S 1957 *The Copernican Revolution* (Cambridge, MA: Harvard University Press)

Landes D S 2000 *Revolution in Time: Clocks and the Making of the Modern World* (Cambridge, MA: Belknap Press of Harvard University Press)

Major F G 1998 *The Quantum Beat: The Physical Principles of Atomic Clocks* (New York: Springer)

McCarthy D D and Seidelmann P K 2009 *Time: from Earth Rotation to Atomic Physics* (Weinheim: Wiley-VCH)

Morris R 1984 *Time's Arrow* (New York: Simon and Schuster)

Muller R A 2016 *Now: The Physics of Time* (New York: WW Norton)

Peterson I 1993 *Newton's Clock: Chaos in the Solar System* (New York: W H Freeman)

Rees M J 1998 *Before the Beginning: our Universe and Others* (Reading MA: Addison Wesley Longman)

Reichenbach H 1991 *The Direction of Time* (Berkeley and Los Angeles: Univ of California Press)

Richards E G 1999 *Mapping Time: The Calendar and its History* (New York: Oxford University Press)

Van Rossum G D 1996 *History of the Hour* (Chicago, IL: The University of Chicago Press)

Smoot G and Davidson K 2007 *Wrinkles in Time: Witness to the Birth of the Universe* (New York: Harper Perennial)

Sobel D 2007 *Longitude: The True Story of a Lone Genius who Solved the Greatest Scientific Problem of his Time* (New York and London: Bloomsbury)

Thorne K 1995 *Black Holes & Time Warps: Einstein's Outrageous Legacy* (New York: WW Norton)

Wald R M 1992 *Space Time and Gravity: the Theory of the Big Bang and Black Holes* (Chicago, IL: University of Chicago Press)

Webb S 1999 *Measuring the Universe: the Cosmological Distance Ladder*(Springer-Praxis Series in Astronomy and Astrophysics) (London: Springer)

Part II – A few journal articles on clocks and time:

Andrewes W J H 2002 A chronicle of timekeeping *Sci. Am.* **287** 76–85

Barrow J D and Webb J K 2005 Inconstant contants *Sci. Am.* **292** 56–63

Bennett M M F *et al* 2002 Huygens's clocks *Proc. R. Soc.* A **458** 563–79

Bensky T J 2010 The longitude problem from the 1700 s to today: an international and general education physics course *Am. J. Phys.* **78** 40–6

Bronowski J 1963 Clock paradox *Sci. Am.* **208** 134–47

Cady W G 1949 Crystals and electricity *Sci. Am.* **18** 46–51

Callender C 2010 Is time an illusion? the concepts of time and change may emerge from a universe that at root is utterly static *Sci. Am.* **302** 58–5

Celnikier L M 1980 Teaching the principles of radioactive-dating and population-growth without calculus *Am. J. Phys.* **48** 211–3

Charette F 2006a Archaeology: High tech from Ancient Greece *Nature* **444** 551–2

Charette F 2006b Archaeology: High tech from Ancient Greece (Erratum 444 pg 551 2006) *Nature* **444** 699

Chou C W *et al* 2010 Optical clocks and relativity *Science* **329** 1630–3

Christianson J 1961 Celestial palace of Tycho Brahe—in late 16th century a danish nobleman built on island of hven most ambitious astronomical observatory world had yet seen—its precise instruments created modern astronomy *Sci. Am.* **204** 119–28

Cranor M B *et al* 2000 A circular twin paradox *Am. J. Phys.* **68** 1016–20

Davies P 2002 That mysterious flow *Sci. Am.* **287** 40–7

Davies P C W 1979 What is time *Sciences-New York* **19** 18–23

Enge P 2004 Retooling the global positioning system *Sci. Am.* **290** 90–7

Feder T 2012 Time for the future *Phys. Today* **65** 28

Fernie J D 2002 Marginalia: Finding out the longitude *Am. Sci.* **90** 412–4

Finkleman D S *et al* 2011 The Future of time: UTC and the leap second earth's clocks have always provided sun time. but will that continue? *Am. Sci.* **99** 312–9

Fischetti M 2004 Rock clock *Sci. Am.* **290** 98–9

Freeth T 2009 Decoding an ancient computer *Sci. Am.* **301** 76–83

Friedt J M and Carry E 2007 Introduction to the quartz tuning fork *Am. J. Phys.* **75** 415–22

Gibbs W W 2002 Ultimate clocks *Sci. Am.* **287** 86–93

Hafele J C and Keating R E 1972a Around-the-world atomic clocks: observed relativistic time gains *Science* **177** 168–70

Hafele J C and Keating R E 1972b Around-the-world atomic clocks: predicted relativistic time gains *Science* **177** 166–8

Hansch T W *et al* 1979 Spectrum of atomic-Hydrogen *Sci. Am.* **240** 94–111

Hartnett J G and Luiten A N 2011 Colloquium: comparison of astrophysical and terrestrial frequency standards *Rev. Mod. Phys.* **83** 1–9

Headrick M V 2002 Origin and evolution of the anchor clock escapement *IEEE Control Syst. Mag.* **22** 41–52

Herring T A 1996 The global positioning system *Sci. Am.* **274** 44–50

Hogan C J 2002 Observing the beginning of time *Am. Sci.* **90** 420–7

Hogan C J *et al* 1999 Surveying space-time with supernovae *Sci. Am.* **280** 46–51

Hurley P M 1949 Radioactivity and time *Sci. Am.* **181** 48–51

Irion R 2004 The pulsar menagerie *Science* **304** 532–3

Kirshner R P 2004 Hubble's diagram and cosmic expansion *Proc. Natl Acad. Sci. USA* **101** 8–13

Lasky R C 2012 Time and the twin paradox *Sci. Am.* **21** 30–3

Lepschy A M *et al* 1992 Feedback-control in ancient water and mechanical clocks *IEEE Trans. Educ.* **35** 3–10

Lombardi M A 2011a First in a series on the evolution of time measurement: celestial flow and mechanical clocks *IEEE Instrum. Meas. Mag.* **14** 45–51

Lombardi M A 2011b The evolution of time measurement. part 2: quartz clocks *IEEE Instrum. Meas. Mag.* **14** 41–9

Lombardi M A 2011c The evolution of time measurement. part 3: atomic clocks *IEEE Instrum. Meas. Mag.* **14** 46–9

Lombardi M A 2012a Recalibration: The evolution of time measurement. part 4: the atomic second *IEEE Instrum. Meas. Mag.* **15** 47–51

Lombardi M A 2012b The evolution of time measurement. part 5: radio controlled clocks *IEEE Instrum. Meas. Mag.* **15** 49–55

Lyons H 1957 Atomic clocks *Sci. Am.* **196** 71–82

Musser G 2010 Could time end? *Sci. Am.* **303** 84–91

Nordtvedt K 1996 From Newton's moon to Einstein's moon *Phys. Today* **49** 26–31

Price D J D 1974 Gears from Greeks—Antikythera mechanism—calendar computer from ca 80 BC *Trans. Am. Phil. Soc.* **64** 1–70

Ramsey N F 1983 History of atomic clocks *J. Res. Natl Bur. Stand* **88** 301–20

Scherr R E *et al* 2001 Student understanding of time in special relativity: simultaneity and reference frames *Am. J. Phys.* **69** S24–35

Scherr R E *et al* 2002 The challenge of changing deeply held student beliefs about the relativity of simultaneity *Am. J. Phys.* **70** 1238–48

Schlenoff D C 2004 A century of Einstein *Sci. Am.* **291** 102–5

Siever R 1975 Earth *Sci. Am.* **233** 82–90

Siever R 1983 The dynamic Earth *Sci. Am.* **249** 46–55

Stix G 2002 Real time *Sci. Am.* **287** 36–9

Tegmark M and Wheeler J A 2001 100 years of quantum mysteries *Sci. Am.* **284** 68–75

Tobin W 1998 Leon Foucault *Sci. Am.* **279** 70–7

Tobin W and Pippard B 1994 Foucault—his pendulum and the rotation of the Earth *Interdiscip. Sci. Rev.* **19** 326–37

Veneziano G 2009 Did time have a beginning? a meeting point for science and philosophy *Two Cultures: Shared Problems* ed E Carafoli, G A Danieli and G O Longo (Milan: Springer), pp 3–12

Weinberg G *et al* 1965 The antikythera shipwreck reconsidered *Trans. Am. Phil. Soc. New Series* **55** 3–48

Wilson C 1968 Kepler's derivation of elliptical path *Isis* **59** 5–25

Wilson C 1972 How did Kepler discover his first 2 laws *Sci. Am.* **226** 92–106

York D 1993 The earliest history of the Earth *Sci. Am.* **268** 90–6

www.ingramcontent.com/pod-product-compliance
Lightning Source LLC
Chambersburg PA
CBHW080550220326

41599CB00032B/6424